Aquilegia pyrenaica

Aquilegia vulgaris

Delphinium Consolida

Delphin. Aiacis

Delphin. orientale

Delph. peregrinum

Delph. Staphysagria

Delph. fissum

Delph. pentagynum

A FAMILY OF FLOWERS

CLEMATIS

AND THE RANUNCULACEAE

DEBORAH KELLAWAY

First published in 1994 by
PAVILION BOOKS LIMITED
26 Upper Ground, London SE1 9PD

Picture credits on page 128

Designed by Janet James

A CIP catalogue record for this book is available from The British Library

ISBN 1 85793 055 X

Typeset by Litho Link Ltd, Welshpool, Powys, Wales
Printed and bound in Great Britain by
Butler and Tanner Ltd, Frome and London

2 4 6 8 10 9 7 5 3 1

This book can be ordered direct from the publisher. Please contact the marketing department. But please try your local bookshop first.

FRONTISPIECE: *Clematis alpina* 'Frances Rivis' heralds the spring.

CONTENTS

PREFACE

The Family

They start coming in the dead of winter, ground-hugging winter aconites, head-hanging Christmas roses; they continue in early spring, carpets of blue anemones self-seeding under shrubs; in April they are showy yellow marsh marigolds in damp places, followed by taller, showier golden globe flowers of May; they are delicate pastel columbines in June, and trembling mauve and yellow thalictrums; and by July they are the glamorous delphiniums soaring up in front of arches of old roses. Then in August they become dark blue monkshoods on the border's edge; in September they are Japanese anemones, pink and white, growing in the shade of apple trees; they are towering, tapering cimicifugas in October. In whatever part of the garden they appear, they are the thing I have been specially waiting for in that particular place, whose non-arrival would be a major gardening disappointment. But month by month, year after year, they do arrive, and often spread and multiply.

The multitudinous flowers of *Clematis* 'Perle d'Azur' mixing with the soft pink *Clematis* 'Comtesse de Bouchaud' outside the author's kitchen door.

They are obviously quite different from each other. Some of them are plump and tiny, others very tall and thin; some have thick shiny leaves and others have soft ferny ones. Most of them are herbaceous plants but two of them are annuals (love-in-a-mist and larkspur) and one of them – the most ubiquitous of all – is a climber.

Clematis – this was the plant that first alerted me to the existence of the family. I was reading J. Fisk's *Success with Clematis*. On page one he wrote:

> They belong to the Ranunculaceae family, of which the buttercup is a member, and this gives us a clue to the likes of the clematis – a moist but well-drained soil.

It was a clue I was not yet equipped to pick up. The Swedish botanist Linnaeus had arranged plants into twenty-four classes in the eighteenth century, but when I used to read a plant catalogue I went for the genus, the species, the variety. The family name might well be there, after the genus and before the species, but I skipped it. It was usually in brackets, anyway, thus:

Genus	**(Family)**	**Species**	**Variety**
ANEMONE	(Ranunculaceae)	*blanda*	'White Splendour'
AQUILEGIA	(Ranunculaceae)	*vulgaris*	'Nivea'
CALTHA	(Ranunculaceae)	*palustris*	'alba'

I pored over catalogues for years and was blind to the fact that the anemone and the thalictrum belonged to the same family. The alphabet had kept them far apart. I certainly was deaf to the extraordinary roll-call of the clematis's relations.

Further on in Mr Fisk's book about clematis, he produced a bit of information which rather shocked me. He said that the coloured petals of the clematis were not, strictly speaking, petals at all, but sepals – segments of the outer covering of the flower when in bud. I felt obscurely that he was downgrading the beauty of the multitudinous, wide-open blue flowers of *C.* 'Perle d'Azur' outside my door (surely sepals should be green, anyway) and thereafter, if I self-consciously experimented with referring to the 'sepals' of the clematis, I felt pedantic. I certainly did not tell my friends that it belonged to 'the buttercup family'.

It was nothing like a buttercup – *Ranunculus repens*, that tenacious, creeping weed which runs along the ground like an embroidery stitch, attaching itself at regular intervals by rooting stems, and making a new plant wherever it touches down. In a moist place it can cover four square metres a year, and its seed is said to survive for forty years in the soil. My sharp trowel fits it exactly: one neat cut, some purposeful leverage, and out it comes entire, its white roots dangling in a circle. It is the most satisfactory of all weeds to dig up. Children, who don't know the difference between weeds and cultivated plants, recognize the charm of buttercups; they hold them under one another's chins; if there is a bright reflection there, that child likes butter.

Charming they may be, but they have a sneaky habit of snuggling up to other plants. Long ago I found a buttercup in a town flower border where an aquilegia used to grow, and someone told me that aquilegias tend to turn into buttercups because the two belong to the same family. This was a piece of *quasi*-botany which even then I found unconvincing, but I resented the relationship of buttercup and columbine, just as I did the grouping of buttercup and clematis under a single head.

Yet classification itself is no problem to us gardeners – it is helpful. We have been learning to classify from the moment we started to learn about gardening. We distinguish trees from shrubs, perennials from annuals, annuals from biennials. We learn to divide plants into hardy, half-hardy and tender, into sun-lovers and shade-lovers, lime-lovers and lime-haters: practical classifications, telling us what to plant and where. We proceed to classify by season (winter flowers, summer flowers) and by colour (with a special compartment for modish white flowers). What we amateurs would not think of doing is to classify plants by their sex-lives; we would not look into the heart of a flower to see whether the stamens were attached below, or above, the pistil. But this is what Linnaeus did.

It was a book by an amateur botanist – John Raven's *A Botanist's Garden* – that turned my mind from its bewildered boredom at the existence of plant families to a sense of pleased discovery. The chapter headings of the book are the names of plant families: Ranunculaceae, Leguminosae,

Rosaceae, Umbelliferae, Euphorbiaceae. The Ranunculaceae come first and the richness of the family was laid before me. There they all were – my favourites: I loved and grew one or two species of almost every genus and half-a-dozen clematis, without knowing that they belonged together.

I turned to the text books. The definitions of the family were confusingly filled with words like 'rarely' and 'usually' and 'often' and 'or':

> Terrestrial perennials (rarely annuals) rarely climbing soft-wooded plants with opposite leaves . . . (J. Hutchinson: *Families of Flowering Plants*).

The paragraph continued to be hedged with qualifications and exceptions, as if the botanist were doing his best to indulge a family of very wayward, independent-minded children.

Occasionally, it has become too much; a plant has had to be kicked out. This has happened to the peony: it used to be among the Ranunculaceae but it has been banished and told to set up on its own under the new name of Paeonaceae. Like all departures from families, it is a loss.

Plant families are like human families: they descend from a common ancestor. The common ancestor of the Ranunculaceae is lost in prehistoric slime, but it lived so long ago that even its beautiful descendants, ornamenting our gardens, are described by systemic botanists as 'primitive'. For this reason the buttercup family is put first in the sequence of plant families, before the Magnoliaceae and the Lauraceae. Again, like human families, the different members of a plant family do not resemble one another in all respects, nor is there one supremely important single characteristic which they must all share in order to qualify for membership. But in the sum total of their characteristics, they must be *more* like one another than they are like the members of any other family, even though none of their separate characteristics is exclusive to their family alone. It is the mixture,

Starry *Anemone blanda* self-seeds amongst primroses and daffodils in the author's woodland garden.

the combination of characteristics, that makes the family distinct.

Each genus shares some features of its nearest relative, and adds some fresh ones. I have heard the whole thing compared to a game of 'Whispers', where everyone sits in a line and words breathed into the ear of someone at the beginning emerge almost unrecognizable at the other end. No wonder the anemone looks different from the thalictrum, sitting at the other end of the line.

But I am going to look for likenesses. Already I can see that the larkspur is a poor relation of the delphinium, and that the monkshood is a rich one. I recognize, with satisfaction, that the globe flower, trollius, looks like a swanky version of the creeping buttercup. I used to think that the basal leaves of the cimicifuga were uncannily like those of the Japanese anemone. Now I realize that it is just a family likeness – not uncanny at all. I begin to discover that these family likenesses are observed in specific names: a thalictrum with blue-green, ferny leaves is called after an aquilegia (*T. aquilegiifolium*), and a ranunculus with dark green, divided leaves is called after an aconitum (*Ranunculus aconitifolium*), while a certain group of aquilegias is called 'clematis-flowered'.

Finding similarities is more fun than finding differences, or at least more reassuring. It brings order to the multifarious world of plants. I cannot yet see where the clematis fits into the family, or how a delphinium or aquilegia could borrow its name. But I shall school myself to look at plants, not just as part of the garden picture, but also in close-up, to follow the family through a single year, from January to December, and watch the succession of 'buttercup' flowers as they open, month by month, in my Norfolk garden.

My garden is about an acre of neutral, sandy, well-drained loam, nicely suited to the Ranunculaceae family if only it were not so dry. There is a herbaceous border, a man-made pond, a spring garden leading to a very small wood, a thatched cottage with white walls, an old outhouse with black ones, fruit trees and four garden arches where clematis can climb.

Each flower has its month. The prodigal clematis belongs to them all: *C. cirrhosa* to January, *C. armandii* to March, *C. alpina* to April, *C. montana*

to May, *C.* 'The President' to June, *C.* 'Victoria' to July, *C. tangutica* to August, *C. jouiniana* to September, *C.* 'Lady Betty Balfour' to October. But there are already candidates for all these months, and only when November and December come are the other flowers spent. Then *Clematis vitalba*, old man's beard, comes to the rescue and completes the calendar, with its seed-head whorls as frothy as white roses along the wintry hedgerows. So I shall save up the genus *Clematis* until the end of the year, and start in January with the winter aconite.

WINTER

THE WINTER ACONITE

Eranthis hyemalis

I rounded a corner of the hedge and suddenly saw them: a dozen tiny yellow globes barely an inch above the earth. I had been feeling disconsolate about the garden. Try as I might, I could not persuade myself that it looked good: it was both untidy and bleak. Wherever I went, rabbits, moles or squirrels had been at work. I was not looking for winter aconites, I thought it was too soon. It was 12 January and the weather had been cold, yet there they were, doing their work, which is to show that the year has turned.

I picked two, digging my fingers into the soil to get enough stem to be plausible, and wondered what vase or vessel I possessed that would be shallow enough to hold them and not swamp them. The answer was a standard-issue eye-bath of deep blue plastic – as blue as the glass lining of an old silver salt-cellar; it was the right size, the right shape and the right colour. Well pleased with the arrangement, I put it on the kitchen table, and sat down to scrutinize the flowers.

A sunny colony of winter aconites.

Close up, they were perfect bright yellow globes (a yellow without a hint of red in it), cradled on sharply scissored collars of leaf-green, 4cm/1½in across, with no neck or stem between the yellow and the green; they seemed inseparably grafted together. The collar, or properly the *involucre*, fell slightly downwards away from the yellow globe, leaving it free to open.

'They will be open tomorrow,' I thought, 'in the warmth of the kitchen.'

But they opened that very evening. If I had sat beside them for two hours, I would have seen them do it. Once open, the flowers were like buttercups, but less lacquered and more reliably cup-like. What I thought were six perfectly rounded petals might have formed the model for a child's stylized drawing of a flower, though the child would draw it flat, not cupped, and would be hard pressed to suggest the central burst of stamens – twenty or thirty of them – fanning out round the pale green pistils like a loosely held bunch. To the naked eye the stamens looked exactly the same yellow as the petals, but under a magnifying glass the anthers were a little more biscuit-coloured, the filaments just a little greener. The total effect remained a unified simplicity of clear yellow and clear green.

Next morning the two flowers were wider open, but I decided to pick two more to fill the eye-bath tight. There had been a frost that night, so white it looked like snow along the hedgetops and on the paving, but now the sun was out and there were the aconites, merry as ever in their sheltered corner, completely unscathed as if nothing had happened. Clearly, the petals were made of stern, weatherproof stuff.

But they were not really petals at all, so the botanists declared. They were coloured sepals. The tiny winter aconite had this in common with the grandly climbing clematis – coloured sepals instead of petals – and I had found the first of the family likenesses I sought.

Two weeks later, aconites were coming up in other, less sheltered corners of the garden. First through the earth came a pale stalk, bent over like a crook and pulling the flower through upside down. Next came the tightly furled involucre. Involucre means 'envelope'; the aconite's involucre is made of three dissected bracts which overlap each other to enfold and protect the bud as it is thrust out into the light. For a day or two the involucre holds close to the sides of the bud and points downwards like the hands of a diver; then it straightens, relaxes, falls back, and the yellow globe shines out.

It will shine from January to March, opening and closing according to the temperature, but never soiled or spoilt. In a sprinkling of snow, when other plants look exhausted and chilled, it will be neat and bright as ever. Indeed, the old herbalists Gerard and Parkinson thought snow improved its performance. Parkinson describes it making its appearance 'in the deep of winter . . . Most commonly after the deep frosts, bearing up many times the snow upon the heads of the leaves.' As the weeks pass, it grows a little taller, though 'not above four fingers high', as he accurately records.

When new batches of freshly picked aconites from my garden proved too tall for the eye-bath, I found a small cream jug, where they looked rather less pretty but still lasted for two weeks or more, opening their six – or occasionally seven – sepals to the warmth. Only when, finally, they lost their freshness and took on a hint of transparency did I bring myself to continue my botanical investigations. I was looking for, and found, 'honey-leaves' or nectaries, lying between stamens and sepals, exactly the same yellow as the sepals, very small and therefore hard to see. But once discovered by the magnifying glass, they seemed quite visible without it: six of them, cleft at the tips, arranged alternately with the sepals, the aconite's answer to the petal and the lure to whatever winter insect was prepared to fertilize the flower.

For winter aconites are great self-seeders. Around the edge of any favoured colony in early May the seedlings can be seen. In the first year they have no leaves other than a pair of cotyledons, or seed-leaves: tiny frilled replicas of the fine-cut basal leaves which appear round the flower stalks in March. It may be another year or two before the seedlings flower. They start with one flower, the next year two, and gradually a whole bright bunch of flowers, five or six of them, each solitary on top of its own stout stem, will grow from a single tuber. The different sizes of the clumps in a colony brings variety among the uniformity of yellow.

By April, the flowers will be gone. The sepals and stamens will have dropped. The involucre itself will assume the status of a flower, round as a daisy. It grows quite high on the old flower-stalk (4cm/1½in), and spreads out like a showy little parasol with a frilled edge. The parasol's central spike is supplied by the pistils: 'Divers small hornes or cods set together,' Parkinson calls them. A whole bed can be carpeted in these green circles, but they finally turn sallow and unwanted before they disappear. They are out of sight for eight months of the year, which is why they are best planted under deciduous shrubs or trees where the sun shines through bare winter branches and the rain can reach them but where, by May, a canopy of leaves will hide their withering. For the rest of the summer they will be forgotten, and undisturbed in the tree's shadow. Dormant, they will not care if the ground under the leafy tree is dry.

But my main spreads of aconites are in the sun. Once I planned a winter garden in two square beds facing south between walls of yew, and planted aconites there. They were interspersed with *Scilla tubergeniana* – washed-out blue little flowers that can be mistaken, when they first appear, for those chips of milky crockery that sometimes surface in the soil. At first the scillas prospered; the aconites loitered. Then they reversed roles. The scillas dwindled, the aconites multiplied; they began to fill the two beds from side to side with carpet-squares of brilliant yellow between walls of dark green yew.

The planting was quite mistaken – too formal and too hot. I tried half-heartedly to move them when they threatened to take over. But they are cussed little customers. This was where they had decided to be, and this is where they took my breath away when I rounded the corner on that bleak January day. The previous autumn I had mulched the beds with mushroom compost with a view to feeding other things. Never had the aconites looked more beautiful than against a background of darkest brown.

But I am pitting my own strong will against the strong-willed aconites; I am working to make an orthodox colony, not square but shadow-shaped, beneath a medlar tree. The medlar's fallen leaves, a thick mulch of cocoa-colour, look almost as nutritious as mushroom compost or beech leaves. Its branches weep to the ground, making a sort of tent within which the aconites can reign, undisturbed – except by bits of ground ivy, lank cow parsley, and scraggly grass. Medlar leaves come late, so there will be sun and rain when the aconites need it, deep shade for the rest of the year.

Meanwhile they are starting their own colonies in ridiculous places: in the middle of a rhubarb bed, at the foot of a compost heap. As quickly as they appear, I move them to the medlar site. To dig a trowel under a fat clump of aconites in flower, and move it somewhere else, brings instant pleasure. Wherever you place them, they go on flowering that year, and with any luck return the next. If one day there is a rich scattering of yellow globes under that twisted wintry tree it will be the loveliest thing in the January garden and one of the loveliest things in the whole garden year.

Everyone who plants an aconite must envisage colonies. Patience is needed, faith and luck. E.A. Bowles regarded a spread of aconites as the 'test' of whether a garden was 'established' or not. 'Your parvenu architect-planned, and colour-schemed affair', he wrote, 'can seldom include such a fine drift of its cheery yellow faces in their green Toby frills as one may see in the garden of many a parsonage or quiet old grange.'

They are their own best company. It is tempting to try them under a winter-flowering shrub – *Viburnum bodnantense, Hamamelis mollis* or the winter-flowering cherry (*Prunus* × *subhirtella* 'Autumnalis') – but they don't need this extra floral boost. 'Mix them with snowdrops,' some experts say, for the long aconite season easily embraces the short snowdrop season and one might suppose that to mix two adorable early flowers together would double the delight. Curiously, it is not so; because each is so tiny, to make an effect they need to duplicate and reduplicate themselves. The exquisite fragility of the snowdrop makes the sturdy little aconite look rather coarse, while the aconite's plump and sunny yellow is not an especially complimentary foil to the purity of the snowdrop's white and chilly green, which looks far lovelier against old twigs and ivy leaves all by itself in a wood.

Eranthis hyemalis is the winter aconite's botanical name, a fitting name for a plant that flowers from January to March, for *er* means 'spring' and *anthos* means 'flower' in Greek, while *hyemalis* means 'pertaining to winter'. Its earliness is its supreme gift – it used to be called New Year's Gifts in Lincolnshire.

But Parkinson called it, ominously, the 'Winter Wolfesbane' because, like all aconites, it is poisonous. For Gerard, it had a further strange and magic property: it made any passing scorpion feel dull and drowsy. Only a touch of hellebore would restore the scorpion to health again. Yet, two centuries later, Linnaeus thought the winter aconite *was* a hellebore: *Helleborus hyemalis*, he called it. This gives hope of finding more family resemblances as I pass from January to February, and from the winter aconite to the Christmas rose.

It is February already, and the whole of England is suddenly covered in snow. Icicles hang from the thatch, a tender clematis is swooning against a wall; the clipped box balls look like Christmas puddings covered in white sauce which is running down their sides, and the potted myrtle has had to come indoors. But I know the winter aconites are alive beneath a foot of snow, their globes shut very tight, waiting for the thaw.

Eranthis hyemalis: close-up, it is a perfect bright yellow globe, cradled on a sharply scissored collar of leaf-green called the 'involucre'. When the sun shines, it opens.

THE ADONIS FLOWER

Adonis amurensis

Another sort of buttercup was alive beneath the snow. A week after my first aconites appeared in January, I saw that a purposeful-looking knob had pushed through the now-frozen soil at the bottom of the garden.

'What on earth is *that*?' I thought.

At a cursory glance you might mistake it for a particularly substantial winter aconite, but it was not coming out of the earth upside down and its ferny leaves were not bright green. I crouched down and found it was a fat bud with a promise of gold about it, tight-set in a whorl of folded filigree. Then I remembered that, out of curiosity, I had ordered a single *Adonis amurensis* the previous autumn, though I thought that I had planted it somewhere else. What's more, I wanted it somewhere else. The crust of the earth was iron-hard, but below the surface it was soft. Foolhardy, I moved my new adonis, with a generous cube of undisturbed soil all round it and complete with frosted top, and replanted it on the edge of my spring garden where I had intended it to be. It sat there unflinching, a few centimetres above the ground, and certainly it was easy to believe that it was the winter aconite's grand cousin. A week later it was decidedly a little taller and more golden and a second stout bud was pushing through beside the first. It was just beginning to open when the snow came.

In ten days the snow receded and the flower reappeared. Compared to the unscathed aconites, its leading bloom now looked a little sodden, there was a suggestion of brown among the gold. Perhaps this fact was a pointer to a crucial difference between adonis and aconite: the golden petals of the adonis really *were* petals and not the surprising coloured sepals which make the winter aconite so weatherproof. Indeed, the adonis's suggestion of brown came from its five sepals, whose yellow is overlaid with brush strokes of umber.

At a cursory glance, the adonis flower is like a marigold; on closer inspection, it is like an outsize celandine. It is 3cm/1¼in across, and though it is many-petalled, it managed to look more single than double, opening flat to display the free boss of stamens spiralling out round its apple-green centre. Twenty elliptical petals can be counted, with five slightly smaller mouse-coloured sepals behind them. Some of the streaking on the sepals seems to have leaked on to the backs of the outer petals, so that it is not immediately clear where the petals stop and the sepals start. The adonis seems, after all, to share the general tendency of the Ranunculus family towards petaloid sepals.

The word 'gold' is hard to resist when describing the brightness of this winter flower. But it is not the right word. Miss Jekyll grew quite angry on this point. 'Gold is not bright yellow,' she scolded in her *Wood and Garden*. 'When we hear of golden buttercups, we know that it means bright yellow buttercups.' And when I write of golden adonis, I mean bright yellow adonis – it is as yellow as scrambled egg made without milk.

And it is eye-catching. The flower opens before the leaves. First comes one flower, solitary at the end of its strong stalk. Then another stout bud appears, and then another, until there are seven or eight stems, 15cm/6in high, the strongest of which branch with a promise of further buds on their tips, so that finally almost a dozen flowers are open or preparing to open on my brand-new plant. What will it do when its tuber is really established? For it is clump-forming, a plant that settles down and is best left undisturbed. I await double the number of flowers in a year or two's time.

As the flowers unfold, so do the leaves – of a colour much harder to pinpoint than clear yellow. It is certainly not clear green; in catalogues it is called 'bronze', but that seems too metallic for such a soft growth: dull, dark olive green is nearer the mark. That such a bright and simple flower should have such subtle leaves redeems it altogether from the ordinary.

But they are typical ranunculus leaves, I find myself thinking, like the leaves of poppy-flowered anemones. They are arranged alternately on the stem, each leaf divided into three, and each division divided again almost to the base by serrated-edged segments which are like microcosms of a whole leaf. They are soft, complicated, delicate, yet the total effect of the plant is sturdy and half-wild. The leaves start as a tuft just below the flower, but not so near as to form an involucre like the winter aconite's; the flower is mounted above the tuft on a few centimetres of stalk.

There are twenty species of adonis. I write of *Adonis amurensis* 'Fukujukai', which is described by Alan Bloom, an authority on hardy perennials, as the only truly winter-flowering herbaceous plant – apart from the hellebore – that he knows. It is a hybrid raised in Japan by a family of nurserymen for whom, over three generations, hybridizing the Manchurian species *Adonis amurensis* has been a continuing passion. They have bred them in green, white, red, purple – as well as yellow. 'Fukujukai' means 'sea of richness'. In 1964 Monjiro Nakamura, the grandson of the original nurseryman, wrote an article in the RHS *Journal* explaining the sea of richness as the wide spread of adonis flowers on hillsides all over the Japanese islands. In the autumn, the Japanese pot them up in readiness for the Japanese New Year, when they bring them indoors to decorate their houses. I would love to decorate winter rooms with adonis flowers. So far, I have only picked one; like the winter aconite, it lasts well in water, opening wide in the daytime and closing at night.

But its preferred place is not in a pot, but under a tree. Monjiro Nakamura noticed that it flourished exceedingly under the mulberry trees which the Japanese plant in their fields for the breeding of silkworms. The mulberry comes late into leaf. Below it, leaf-mould accumulates. This is what the adonis likes. A spot under a deciduous tree is right for it, just as it is right for the winter aconite. Failing a mulberry, a cherry, a magnolia or – for a spectacular harmony of yellow – a *Hamamelis mollis* will do.

By July, the plant will have disappeared. Again like the aconite, it is dormant for six months of the year, easily forgotten and easily spiked with a garden fork if placed in a conventional border. Mine now finds itself in a small open space between a spreading *Viburnum plicatum* 'Lanarth' and an about-to-spread *Hydrangea villosa*: not an inspired placing, but serviceable, for there is loamy leaf-mould, moisture, sunshine in the spring and shadow in summer. Near by is a young pear tree. I shall plant my second adonis underneath it.

The second adonis is *Adonis amurensis* 'Flore Pleno'. The very name sounds a warning: it is double, and in my book 'double' means 'half as attractive'. Also, it begins to flower a month later than *A. amurensis* 'Fukujukai', which takes away another half of its value. Lest this sounds as if we are left with no value at all, we are in fact left with a cheerful bright yellow flower with a pale green

ABOVE: *Adonis vernalis* flowers in spring, lemon-yellow with the typical free stamens of the buttercup family.

RIGHT: *Adonis amurensis* is like an outsize celandine with deep olive-green, divided and serrated leaves.

centre; this green centre is its claim to inclusion in the garden.

There is another perennial adonis, *A. vernalis*, flowering (as its name suggests) in spring, and some gardeners think it the best of the bunch: not bright yellow, but primrose; not in the least double, but demurely single. Yet only a handful of nurserymen supply it in the current edition of *The Plant Finder*.

It is the same story with all adonis species. Everyone agrees that the perennial varieties are robust, easy-to-grow, long-lived, useful. They enjoy lime but perform well on acid soil, too. Yet if you walk into a plant centre in the first two or three months of the year, when flowers are scarce and a splash of yellow would uplift the herbaceous decks, you are unlikely to see one. The reason may lie in the relative difficulty of propagation. You cannot, for example, raise *A. amurensis* 'Fukujukai' from seed, for it is sterile. Most species do not enjoy division, and take a time to re-establish themselves. None of this matters in the garden once you have secured your specimen. You need do nothing but sit back and watch it thicken, perhaps feeding it with bonemeal or fishmeal between autumn and early spring when its root system is growing, and scattering ashes round it in spring to keep the slugs away.

There are annual adonises, too, and once upon a time, we are told, they used to spread abundantly in English arable chalk-land and were even sold in bunches in Covent Garden. 'Pheasant 's eye', they

were called, because of the dark splash at the base of each petal giving a black pupil to the eye of the flower. The flowers themselves were deep scarlet-red: 'The red flowers of Adonis,' wrote Gerard in 1597, 'groweth wilde in the west parts of England among their corne.' He was referring to *Adonis aestivalis*, the summer annual, yet it is barely listed in the proliferating wildflower catalogues of today, though Suttons supply seed that proves beautifully easy to sow: pale enough to show up against the earth, large enough to sprinkle thinly with the fingers. Perhaps it is about to return and will presently be seen along the fringes of cornfields, less invasive than poppies, a deeper red, and the bearer of a much grander name.

Adonis, the beautiful youth beloved of Aphrodite, was gored by a boar. Wherever a drop of his blood fell, so the story goes, a beautiful red flower sprang up. It was *flos adonis* – the adonis flower, although often the books called it an anemone instead. Perhaps the myth-makers thought an adonis *was* an anemone; family likenesses breed confusions.

My yellow adonis clearly did not spring from Adonis's blood, but in its herbaceous habit it observes the central mystery of the Adonis cult: the idea of death and resurrection. When the wild boar gored Adonis, he seemed to die. He sank down into the Underworld for six months of the year; but for the other six months he returned to earth, heralding the spring. *Adonis amurensis* 'Fukujukai' does the same. From July to December, it visits the Underworld. But each new year those fat, astonishing buds come pushing through the wintry earth.

HELLEBORES

Helleborus

Christmas roses are reluctant to flower at Christmas. Mine were not seriously noticeable until February, though long before that the little buds could be seen sitting on the surface of the earth like pale snail-shells in the centre of an untidy spread of last year's old rusty leaves. There were two things to be done: first, to put a high cloche over the plant to protect the flowers, when they did open, from splashes and perhaps persuade them to grow taller and flower more punctually; second, to cut off all those prostrate, wide-sprawling, tarnished leaves which had nothing more to contribute to the health of the plant – it looked much better without them, and the cloche rested more comfortably on the ground. But cloches contribute nothing to the look of the garden, either. Mine are cheap plastic eyesores, only to be tolerated for the sake of the buds they protect. These buds gradually grew, on strong stalks, to about 20cm/8in, when they were ready to be cut and brought indoors.

Christmas roses are for picking. Because they look so much nicer in a vase than under a cloche, there need be no compunction about robbing the garden of its choice early flowers. I brought them in as soon as they opened, with a minimum of dallying between flower-bed and house. There was a pin at the ready just inside the back door. I pricked each one at frequent intervals all the way up each pale and juicy stem, then plunged them up to the necks in water for an hour or two before arranging them. Some experts advise dipping the ends for a second or two into boiling water and then cutting a slit halfway up the steam. Whatever technique you adopt, they need water – fast. Then, if you have caught them in a good mood, they may live for days – even weeks – far out-distancing the snowdrops and *Iris unguicularis* which were their companions in my February vase. But if they are in a bad mood and start to wilt, they can be startlingly revived by cutting the ends again beneath the water level. They need careful placing in a vase; they never look up towards the ceiling; they have a tendency to look down towards the floor, and need tactful persuasion to face directly out into the room. If you put the vase on a window-sill, the white becomes translucent – the light is diffused through it as through a frosted light-bulb.

They are magically beautiful, a mixture of ivory-white, pale primrose-yellow and apple-green. The translucent white comes in the five wide-spreading 'petals' which (it is no great surprise to learn) are really petaloid sepals: strong, long-lasting, frost-proof, in texture not unlike white kid. Each one is smudged at the base with apple-green, and against the green centre thus formed there explodes a sparkler of stamens – perhaps fifty or a hundred of them, each with a slender white filament topped by a large, soft, primrose-yellow anther. When the bud first opens, these stamens are tight-packed together in a solid, domed powder-puff; as the flower matures, they separate and fan freely outwards. Between the boss of anthers and the sepals there are eight to twelve nectaries, exactly the same green as the smudged centre of the flower which makes them hard to see, but once identified, they are clearly visible without the aid of a magnifying glass. There is no escaping the family

ABOVE. *Helleborus niger*: the 'petals' are really sepals, long-lasting, frost-proof, in texture not unlike white kid, each one smudged at the base with apple-green.

LEFT: *Helleborus orientalis* begins to flower in February, holding its flowers high on its stems. Unnamed seedlings open in rose, green, white, mother-of-pearl pink and plum.

likeness here: the structure of *Helleborus niger* repeats the structure of *Eranthis hyemalis*: coloured sepals, funnel-shaped nectaries, a multitude of free stamens and, in the very centre, three to five slender green styles with pointed tips. Parkinson noticed the similarity long before Linnaeus grouped them together; he described the flower as:

> like unto a great white single Rose . . . with many pale yellow thremmes in the middle, standing about a green head, which after groweth to have divers cods set together, pointed at the ends like hornes, somewhat like the seed vessels of the Aconitum hyemale, but greater and thicker.

The whole flower is indeed 'greater and thicker' than the aconite: it is at least three times as large. It has no involucre, only a pair of undivided lime-green bracts on the stem behind the flower. Parkinson may have thought them like a great white single rose, but the sepals of my ordinary *Helleborus niger* are pointed, meet almost edge to edge, and when they are wide open they are less like a rose than five-pointed star.

There are grand named varieties of *Helleborus niger*. I have a particularly famous beauty called 'Potter's Wheel'. Its name suggests its circular smoothness. Its sepals overlap each other; they are rounded, not pointed; some of them develop rose-pink flushes on the outside and when the buds are half-open they are tulip-shaped. The open flowers are thrillingly large, on tall stalks, and there are often two flowers to a stem, one growing from the axis of each bract. But in the end they are not always more beautiful than many unnamed forms which have particular sharpness of outline and simplicity.

The Christmas rose is a paradoxical flower; it is white, yet it is called black – *Helleborus niger*, the black hellebore – because its rhizomatous rootstock is encased in a blackish-brown skin. The flower is cool and pure, the root deadly poisonous; both men and beasts are said to have died of it, though it was used medicinally for centuries. Gerard prescribed it as a purgative for 'mad and furious men'. Gilbert White of Selborne reported that its powdered leaves could be given to children 'troubled with worms'. It heralds the spring, but there is nothing spring-like about it; the whole plant is leathery, sculptured, timeless. The Elizabethan poet Spenser listed it ominously among the flowers growing in the unearthly garden of Proserpina. In the nineteenth century it inspired one German romantic poet, Eduard Mörike, to write sepulchral verse (later set to music by Hugo Wolf). He first encountered it in a churchyard. It made a dramatic impact on him; he hymned it as a strange flower of the moon, not the sun, of coldness, not warmth. He detected a barely perceptible fragrance in its heart; it filled him with yearning and, as well as turning it into a mystical poetic symbol, he dug it up and took it home and planted it in his window-box. Thence it was blown away in a high wind.

But it would not have lived, anyway. If Mörike could have read Brian Mathew's Alpine Garden Society monograph, *Hellebores*, he would have learnt that it is inadvisable to dig up whole clumps of *Helleborus niger* and transplant them intact. The best you can do is break them up into small divisions, each with its own leaf-shoot, and plant these divisions in wide holes filled deep with compost and leaf-soil. Gertrude Jekyll used to do this, digging up her Christmas roses in spring, 'washing out' the clumps and then performing delicate surgery with a sharp knife at what she called the 'points of attachment'. But herbaceous hellebores are long-lived, if undisturbed; I am well content to leave mine where they are, feeding them with a general fertilizer and mulching them generously in spring, then watching them slowly thicken and prosper. By February the cloches are off and the white flowers are coming thick and fast and more and more and more, like Lewis Carroll's oysters scuttling up the beach. If you cut one stem it makes more room for another.

There is no problem about buying hellebores – except in paying for them. They are top-favourites and the garden centres are full of them in February. It is a good time to buy them, when they are in flower and you can see what shape and colour you are choosing. The pots of *Helleborus niger* will look full of promise, plump and clean; but beside them or beyond them will be a different range of hellebores: dusky beauties in pink, plum and purple, brushed with livid green. These are specimens of *Helleborus orientalis*, the Lenten rose.

Helleborus orientalis is an easy plant to grow. It will survive on heavy clay or chalk, though what it prefers – like all its family – is leaf-mould and moist loam. It is a gift to town gardeners for, like *H. niger*, it will prosper in deep shade. I have seen it planted along a cheerless passageway between the side of a semi-detached house and a boundary wall; the sunless concrete tunnel was transformed in February by these sophisticated, strange flowers that look so difficult and are so easy.

The label 'Lenten rose', like 'Christmas rose', is misleading. *H. orientalis* starts to send up its strong flowering stalks in early January and is in full flower in February, just like *H. niger*. It is considerably more showy, a garden plant rather than a plant for a vase. While *H. niger*'s flowers seem to jostle one another near the ground, *H. orientalis* holds its flowers distinct and high, three to a 20cm/8in stem, elegantly balanced. If it is happy, it may start to seed itself, producing progeny in shaded pinks and greens.

It is largely unnamed seedlings that the nurserymen will offer; named varieties are collector's pieces, priced to match. There is a creamy subspecies with an exotic, purple-spotted throat: *H. orientalis guttatus* (*guttatus* means 'spotted'). Then there is the Ballard Strain, named after Helen Ballard who has refined *H. orientalis* towards perfection, sometimes crossing it with an ominously dark species from the Balkans called *H. torquatus*, and rejecting all but the very smokiest blue-blacks, the subtlest greens and pinks – she has even produced lime-yellows, and she expects her plants to hold their heads high. A specialist grower explained to me that a Ballard hellebore is the Rolls-Royce on the hellebore highway. But unless you are a keen collector, you will find a nameless Lenten rose a vehicle of beauty.

Mine are nameless, old-rose coloured, shading to lettuce-green at the eye. They hang their heads. The saucer-shaped flowers are darker on the outside than the inside of the sepals. But if you turn one of these saucers up to the light, you will see that its structure is the same as that of the Christmas rose: it is 5cm/2in wide, and there is the same explosion of pale stamens – only here the nectaries show up more clearly: flat, shining green pouches immediately visible against dusky pink; you can even see a sort of division, like lips, at the tip of each pouch. Does this exude the honey? For nectaries are sometimes called honey-petals, or honey-leaves.

The true leaves of *H. orientalis* come from the base of the plant, like those of *H. niger*. They are tripartite, pedate (bird's-foot shaped). They, too, need to be cut away in winter, or they may encourage a fungus disease called 'black spot' which spreads in blotches from old leaves to fresh flowers, threatening complete collapse and calling for urgent and repeated doses of systemic fungicide. All being well, however, new, shining bright green leaves will come after the flowers and will spread out to form ground-cover in shady corners through the summer.

Long ago, botanists used to divide hellebores into two main groups: herbaceous hellebores, like *H. niger* and *H. orientalis*, with flowers and leaves on separate stalks, and evergreen hellebores like *H. foetidus* and *H. argutifolius*, with flowers and leaves sharing the same woody stem, persisting from the previous year. Nowadays this simple division is found to be inadequate. The classification of hellebores is a botanist's nightmare of species, subspecies and infinitely complicated hybrids. And then there are mistakes and confusions in nomenclature. For instance, the very earliest hellebore to flower, *H. atrorubens* of gardens, deep purple and reliably open by Christmas (raised among many others by the famous old firm of Barr and Sons), is actually a hybrid between

H. orientalis and a species called *H. atrorubens*, also deep purple, growing wild in the mountains of Croatia. In Sir Frederick Stern's garden at Highdown, where all manner of hellebores flourished on chalk, strange and beautiful intermediates sprang up in shades of green and cream touched with plum – hybrids between *H. argutifolius* and *H. lividus*. 'The Kew people', as he called them, named one of these hybrids *H.* × *sternii*; it produced seeds, and now has sought-after descendents: 'Broughton Beauty', raised by Valerie Finnis, rather like a refined pink *H. argutifolius*, and the Blackthorn Strain, in exquisite jade-green with prune smudges on the backs of the sepals and marbled leaves, raised by Blackthorn nurseries. Hellebores are erratically promiscuous. *H. niger*, for example, is reluctant to hybridize with its closest relative *H. orientalis*, but happy to cross with *H. argutifolius* or *H. lividus*.

Out of the fifteen or so distinct species of hellebore, I grow a modest four. I eschew the reportedly beautiful *H. lividus* because it is too tender, and the native green hellebore *H. viridus* because its flower is too small. But two evergreens I must have, as a foil to *H. orientalis* and *H. niger: H. foetidus* (the stinking hellebore) and *H. argutifolius* (which for years everyone called *H. corsicus*).

H. foetidus is said to be easy – as easy as *H. orientalis* and as ready to seed itself. I bought one; it prospered; its leaves were remarkable, leaden-green, curved and razor-thin, radiating from a centre like the spokes of an umbrella. In its second year it produced an opulent, loosely branched flower-head from which hung clusters of little apple-green globes, two to a stem. I picked them and found that, contrary to expectation, they lasted in water (laced with a splash of gin). In its third year it collapsed and died. It is the only

ABOVE: *Helleborus* × *sternii* 'Blackthorn' strain.

RIGHT: *Helleborus foetidus*: the leaden-green leaves, curved and razor-thin, grow on the same thick stems as the clusters of pale green flowers and globular, pendant buds. They last for weeks, and some varieties smell sweet.

hellebore to have a weakly formed rhizome, I read. I bought another very small and precious one called 'Miss Jekyll', which promised not to stink but to smell sweet. The same thing happened: it grew healthily, prodigiously, until large enough to flower with elegance; the next year it was dead, its two stems and lovely inflorescence lying limp and rubbery along the ground. Gritting my teeth, I have bought a third one, a swanky cultivar called 'Wester Flisk' with claret-red leaf-stalks and central stem. I imagined its name to come from some Scandinavian horticulturalist, but discovered it is named after a Fifeshire village on the Firth of Tay. I am now looking forward to its flowering and dreading its demise.

Helleborus argutifolius (*H. corsicus*) is another matter, robust and vital, prepared to grow anywhere, in sun or shade, on acid clay or chalk, in beautifully prepared loam or, self-seeded, in the mortar of a wall. It would be flamboyant except that it is so cool and green. In its native Corsica, wild clumps are a metre or two wide, with up to fifty woody stems rising from the base, each crowned by an inflorescence of pale green bracts. The bracts curve backwards to reveal globular, pendent buds; the buds open into hellebore flowers which, when fertilized, turn up towards the sky, at which point they reminded Vita Sackville-West of small green waterlilies. There is absolute unity of colour: stem, bract, bud and flower share an identical apple-green – the green of young beech leaves in spring. Below them, the darker evergreen leaves are tripartite, held in stiff curves, surfaced like soft niger leather, each leaflet sharply toothed along the edges (*argutifolius* means with 'toothed' leaves).

In my garden, seedling plants given by a friend three years ago already spread wide with at least twelve noble stems, 120cm/4ft tall. When the Norfolk wind blows, the outer stems tend to capsize – an undignified thing to happen to such a stately plant. It continues to flower, however, leaning on its elbows, and looks far better thus than tied upright to a stake.

Wherever *H. argutifolius* is planted, or plants itself (for it is a great self-seeder) it will look good. Whatever its neighbours, they will seem to enhance it. If an old seed-head of honesty has blown in among it, the silver will look so pretty entangled with the green that the gardener may be slow to tidy it away. My plants grow on either side of a grassy path leading down to a dry ditch. This turns out to work well – not on the way down, but on the way back up the slope, when you can see the nodding flowers from below. On one side they hang against the branches of the red-stemmed dogwood, *Cornus alba* 'Westonbirt'. The bare stems of mahogany red are so complementary to the hellebore's greens you might think (erroneously) the effect was calculated. On the other side, primroses cluster – and again you might think it a planned harmony, for the green of the hellebore is touched with yellow, and the yellow of the primrose is touched with green.

Everybody who writes about hellebores has different advice on how to place them. One person likes to grow them under the winter-flowering cherry (*Prunus* × *subhirtella* 'Autumnalis'); another mixes them with ferns; someone recommends blue pulmonaria with white Christmas roses, and white pulmonaria with pink Lenten roses. Someone else, preferring pink against pink, plants *H. orientalis* hybrids in front of the purple-pink flowers of *Daphne mezereum*. John Raven fitted them into a border among old roses, lilies and 'sundry onions'. Gertrude Jekyll 'found it convenient' to grow them among species peonies. 'They are agreed in their liking for deeply worked ground,' she wrote, 'with an admixture of loam and lime, for shelter, and for rich feeding.' I have seen the same mixture of hellebores and peonies in

Mrs Merton's Berkshire garden, The Old Rectory. It is more than horticulturally 'convenient', it is visually inspired; the young peony shoots belong to the same colour range as the *H. orientalis* hybrids, where the deep reds and dusty plums, the purples and the indigos are never harsh, but rich like vegetable dyes in Persian rugs. Even lovelier is the effect among shining red peony stems of all the pale hellebore hybrids, with greens the colour of shallow sea-water over sand, and whites tinged with pink like mother-of-pearl.

Only two backgrounds fail to please the hellebores: one is supplied by coarse old elephant's ears – bergenia – flowering in a strident pink which shouts down the hellebore's subtleties; and the other is bare earth, or a mulch of milk-chocolate coloured leaf-mould, against which *Helleborus orientalis* fails to show up at all.

Mulch thickly and then plant thickly – that is the solution. If need be, plant all the hellebore species close together; the green of the evergreen sorts picks up the green in the eye of the herbaceous sorts and harmony reigns; snowdrops will serve to carpet the ground – or anemones.

As winter ends, I have come upon a chance meeting of blue anemones (*A. blanda*) and dusky pink hellebores in my garden and resolve to plant all future hellebores where anemones are spreading. The hellebore flowers dangle at just the right height above the anemones' upturned faces, and there is a hint of pink in anemone blue. But now I realize, with disappointment, that the idea is no good. Throughout February the anemones will not be there and the hellebore will seem to be surrounded by the very thing I most want to avoid: bare earth. Anemones belong to spring.

SPRING

EARLY-FLOWERING ANEMONES

Anemone blanda and others

The first *Anemone blanda* opened, this year, on 7 March. Two days later, because the sun was shining and there was little wind, warmth opened dozens more. The leaves must have been there for a little while; they sneak through the ground when no one is looking, and you don't notice them until, suddenly, a flower opens. Before that you fear that perhaps they won't be there at all this year; perhaps you have carelessly dug them up. When they first come through the leaves have a sort of protective coloration of pinky-brown, almost invisible against the leaf-mould that surrounds them, and they lie flat upon it, hugging the ground. Only when you look close do you see them, and observe how fine-cut and lacy they are. Later, the leaves grow a little higher, and take on a sober green, still touched with dull red which lingers on the stem. They spread out, softly overlapping each other and finally establishing rounded domes of foliage 15cm/6in tall as a foundation for the flowers, which rise on slender stalks 2–5cm/1–2in above

The tousled seedheads of *Pulsatilla vulgaris*.

place, would be enough to delight the eye; it does not need multiplication to make its point. And yet that authority on anemones, E.A. Bowles, disparaged it after Robinson had shown it to him in his garden. Bowles did not forgive it for promptly closing whenever the day was dull, and he mocked its closed sepals for showing what he called their 'cotton backs'. This does not put me off my determination to possess it. Meanwhile, I have one tiny variant of the wood anemone, *A. nemorosa* 'Vestal', with flowers as small as a pea tight-packed with white anthers pretending to be petals.

My second woodlander is *A. ranunculoides*, which botanists have now renamed *A.* × *lipsiensis*. Its earlier name helpfully suggests its appearance, for it is small and yellow and reminds people of buttercups. It also has a look of the celandine about it. But it reminds me most of the winter aconite. Its yellow cup is framed by a whorl of leaves like the aconite's involucre – not so much ferny as palmate: starry foils to the flowers which, like all anemones, are held with delicate poise on the same stem as the leaves but a little way above them. It is bad luck on *A. ranunculoides* that its own identity must often be overlooked – nobody likes being confused with their relations. But it is an unassuming little plant, a carpenter, not a star performer. Like *A. nemorosa*, its rhizomes will spread horizontally and will very gradually take over empty corners, covering them with a dense, soft spread of anemone leaves crowned with countless pure yellow, five-sepalled flowers.

My third woodlander is *A. sylvestris*, the 'snowdrop anemone', so called because, until it is fully open, it droops its head on the top of its tall 30cm/12in stalk. It is satiny white, and scented, and Reginald Farrer thought it one of the loveliest anemones of all. It carries the anemone season on from spring into summer; it can be in flower from April to June. And when it opens, its flower is a little replica of the Japanese anemone of September borders.

Meanwhile, at Easter, comes the Pasque flower (if Easter falls in April). It used to be called *Anemone pulsatilla*; now it has been reclassified as *Pulsatilla vulgaris* and we are not allowed to call it an anemone any more, only to say that it is closely related to the anemone. And so it is – the merest glance confirms this. When its elegantly pointed buds open you can peer into the deep cup of the flower, turned towards the sun, and rediscover the familiar pattern: six coloured sepals round a full boss of stamens, the sepals being paler outside than in, which makes a glimpse inside exciting, as when someone throws back a coat to reveal a rich lining – parma violet, amethyst or ruby, jewel colours set round a golden centre. The plant is tousled and fly-away, wild, outlandish and ravishing. The leaves are as fine-cut, as 'snipt and jagged', as any anemone's, but their dull green is silvered, hairy, and so is the stem. Silver silkiness and curliness set it apart from its relations. In bud, the flower-stems curve over and there is a stiff, cone-shaped involucre, frosted and fringed, protecting the flower. And when flowering is over and the sepals fall, the seed-heads remain, rising from the green pincushion, spidery and faintly sinister, yet silky as everything else about the plant and shining like newly shampooed hair.

In nature, it grows wild on chalk downs in full sun. I used to think my garden soil was too neutral for it; now I am filled with hope that it will

Anemone sylvestris flowers between April and June. Its likeness to the familiar Japanese anemone is striking, but it is shorter, earlier, scented and has an exquisite fragility.

flourish in very narrow borders cut between grass and the concrete slab which marks the septic tank. Perhaps the soil there will be moist but well-drained; perhaps the concrete slab will introduce a taste of lime; perhaps, next year, my Pasque flowers will form substantial clumps like those I envy in other people's gardens – clumps crowned with 30cm/12in flower stems, and as many as twelve flower stems to a single plant. They will not be mixed with other things, but only with variants of themselves in white and Persian red. The Pasque flower is too original in appearance to be an easy mixer.

Yet I have planted one of its small relatives at its feet – another reject from the genus anemone: the little blue hepatica. Once it was *Anemone hepatica*, now it is *Hepatica nobilis*. While my faith in the Pasque flowers burgeons, I allow myself almost no faith in the hepatica's future; already the rabbits have eaten it once, though it has bravely produced new, shining, three-lobed leaves again. These leaves, reminiscent of the three-lobed liver, are responsible for the hepatica's offputting name. It is a tricky little plant, an enchantress which turns out to be a deceiver. Its china-blue anemone-face, shining from a professional grower's pot, tempts one to buy, but once planted in the garden it tends to disappear from the shady spots it is supposed to like. That is why I am now trying it near the Pasque flowers in the sun, where its roots can find their way under the concrete slab.

There is one last group of spring anemones which I cannot imagine in my garden. They are the poppy anemones *A. coronaria* and *A. fulgens*, not altogether hardy or long-lived, but familiar above all other anemones in florists' windows. When you see them in winter the flowers are usually folded inside their involucres which are as elaborate as the pulsatilla's, and if you give a bunch to someone you may feel quite ashamed of them, tight-shut and almost dead-looking. Their subsequent opening in water is miraculous: Colette thought it was 'like a parachute seized by a gust of wind'. Purples, brick-reds, pinks, deep mauves, maroons and whites unfold and there, in the centre of each, are the free stamens, inky blue-black as the centre of an oriental poppy.

These are the flowers, in their scarlet dress, which account for the anemone's name. Reginald Farrer pointed out that the Syrian cry '*Na-ma'am!*' (a cry of lamentation over the dead Adonis) was the origin of the word: a-ne-mon-e. There is only a minor problem here for followers of the family Ranunculaceae, who already believe that the adonis, and not the anemone, was the flower that symbolized the death and rebirth of the beautiful youth. It is annoying to find *two* flowers symbolizing the same drops of mythical blood, even if they do belong to the same plant family. But if my suspicion that the adonis flower was once called an anemone is correct, the confusion lessens.

Anemos means 'wind'. *Anemone* means, literally, 'daughter of the wind'. Pliny believed that anemones only open when the wind blows on them and so should inhabit windy places. My experience is the opposite. This year it was a gentle March sun which made my anemones open, and a cold April wind which is finally blowing the delicate blue sepals away.

Marsh Marigolds

Caltha palustris

If you haven't got a pond – or better still, a marshy stream at the bottom of your garden – you might as well forget it. It is *Caltha palustris*, the marsh marigold. *Palustris* means 'marsh'. It does not mean 'dry ditch', which is where I originally tried to grow it. It was brave, and had the usual will-to-live of flowering plants, so it hung on through its first spring, very sweet and rather dwarfed, but failed to reappear. Our ditch runs with water in wet winters, but dries out in summer.

The garden centres mislead us in this. Marsh marigolds take their alphabetical place in nice, well-watered pots among quite ordinary, dry-land plants beginning with 'C' – *Campanula*, for instance, or *Coreopsis* – and you think you can get away with them in your flower borders with the help of a watering can. The instructions on the back of the plant label do not help: 'Easy and reliable', they say, 'in any soil which does not dry out'. But soil that does not dry out is mud, and it is mud that the marsh marigold prefers; its natural home is a bog or the wet margins of a stream. We have a concrete garden pond and the beds round it are dry, but it is shallow at either end, so we finally put marsh marigolds in the shallows of the pond, in large pots filled with loam and balanced on bricks, juggling heights until the water came level with the surface of the soil. They rewarded us by growing; within two weeks the glossy bright green leaves had spread outwards and the rims of the pots disappeared from view.

It was the white variety, *Caltha palustris alba*, that I vainly searched for in our rough ditch this spring and sorely missed. In the end I acquired a new one, beautifully pot-grown, and put it, temporarily, on the kitchen table where the winter aconites had stood three months before. Here again was a simple flower-face, composed of five rounded sepals with a central boss of yellow stamens. There was roundness everywhere, in leaves and buds as well as flowers. The sepals looked as if they were made of wax, some with a brush-stroke of mushroom-pink on the back, matching the colour of the stalks. The stalks were straight and strong and branching; the flowers grew from the axils, sometimes one, sometimes two, so perfectly presented they might have been wired by a florist on to a bridesmaid's posy. They opened flat, each sepal separate from its neighbour so that you could see space between. The colour scheme was roughly the same as that of the Christmas rose: white, yellow and green, but the effect was entirely different, less pure and chill. There was no green shading at the base of the sepals; the bright green came in the rounded leaves. If *Helleborus niger* belonged to the moon, this artless flower, wide-open and generous, belonged to the sun.

The pond was just stirring into life in early April when we placed it there on its submerged island of bricks; the water iris were barely a hand's span tall, the waterlily leaves only just breaking the water, the fish half-dazed, lying motionless near the surface to shake off the torpor of winter. And now there were white flowers at the end of the pond. They went on coming; the first wave of sepals fell, leaving feather-dusters of stamens; the stamens fell, leaving stiff, shining bunches of green carpels, eight to a flower, hooked at the ends; but reinforcements were rising from the mud – globular buds preparing to take over. They went on coming until May; then, during the summer, the leaves grew much bigger and made a luscious mound of green, but in autumn we were promised a final flush of flowers. Compared with most water plants, *Caltha palustris alba* remains within bounds and for this, as well as for its beauty, it wins my Marsh Marigold Prize.

The popular *C. palustris* 'Plena' (or 'Flore Pleno') is not, in my view, a prize-winner, though absurdly floriferous: it fails both on shape and on colour. Even Gertrude Jekyll might have allowed that the colour is *gold*, if not positively orange. *Plena* means 'full'; the flowers are as double as a pompon dahlia. Surely, one would think, these are petals, held within the five golden sepals that form the outer cup? But they are not. Not content with coloured sepals masquerading as petals, this flower has coloured stamens masquerading as sepals: layer upon scalloped layer of them, supplying the impression of petalled fullness while in fact the flower has no true petals at all. Under a magnifying glass one can accept the botanist's word that these are extraordinarily blown-up and fattened versions of the lavish stamens one finds in the centre of every buttercup flower. Barely perceptible is the stigma, a green eye half-hidden in the midst of the flower's virtuoso performance. One might well imagine it to be the result of the modern hybridizer's skill, but in fact it has been known in England since around 1600, and Parkinson judged it the only marsh marigold worthy of a place in his garden. Some of its forms are sometimes prefixed with the word 'Monstrosa' (as in 'Monstrosa Plena'), suggesting, perhaps, its departure from the norm. Aware now of its oddity, I have a grudging interest in it, and have plunged it into the pond beside *C. palustris alba*, where only a glance is enough to confirm that you do not have to pile on the sepaloid stamens to be pretty: five white, waxy sepals win every time.

But they get on well together, these two marsh marigolds, side by side in the pond. Their leaves are almost indistiguishable from one another, except that there is slight toothing along *C. palustris* 'Plena's' edges; caltha leaves are not divided and dissected like those of most Ranunculaceae, but heart-shaped, rounded like the leaves of celandines. The branching stems take oblique angles, stretching out sideways so that gold and white flowers intermingle. This *C. palustris alba* tones down *C. palustris* 'Plena's' tendency towards vulgarity; *alba* would look better alone, but 'Plena' behaves better in the company of its close relative than it would if seeking attention all by itself. It also behaves beautifully beside a blue cloud of the water forget-me-not, *Myosotis palustris*, and that is how I would like to grow it next year. I shall certainly not grow it beside either of my single yellow marsh marigolds, *C. palustris* or *C. polypetala*: single and double yellows in juxtaposition only minimize one another.

Caltha polypetala has no need of neighbours; strong and adventurous, it is happy to grow, not just in mud, but in water, spreading sideways by stolons, rooting as it goes along. Its stems are as pink as rhubarb sticks, its leaves are as large as a waterlily's, though not as flat – more folded than floating: big leaves for a big pond. Its earliest flowers, opening in April, can be as much as 6cm/ 2½in across: E.A. Bowles thought it the 'emperor'

Caltha palustris, the kingcup of marshes and damp meadows, is the beautiful April wild flower that has been beloved in literature for centuries.

of its genus. There is a story that it originated in a Vatican pond, whence a keen horticulturalist hooked it out with his umbrella while the attention of the custodians was distracted by his aunts, but Reginald Farrer, who told the story, also dismissed it as untrue. Its specific name of *polypetala* is a howler: far from being many-petalled, it has, of course, no petals at all, though it is not always content with five sepals, and can occasionally produce seven, or even ten (though it is not for this reason that botanists have renamed it *C. palustris palustris*). I am watching my *Caltha polypetala*: if it begins to usurp the waterlilies' space it will have to be pulled out, in handfuls. But I am also watching it with hope and pride, waiting for its flowers; those flowers just escape being coarse through their joyous opulence.

My final member of the genus *Caltha* is the simple species itself, *C. palustris*. It is the big, beautiful wild flower which has been well-beloved in the European countryside for so long that it has collected strings of friendly and teasing pet-names like Hobble Gobble, Water Dragon and Soldier's Buttons. It has ninety English common names, as well as sixty French and over a hundred German. It is the 'Marybuds' of Shakespeare's 'Hark! hark! the lark' and it has other Christian names: 'Marygold' suggests dedication to the Virgin Mary rather than likeness to the marigold (which belongs to the daisy family). But its favourite English common name is kingcup.

Lucky people come upon kingcups growing wild in damp meadows or beside streams. I came upon them this spring on a fen – brilliant in the April sunshine among last year's dried Norfolk reeds and this year's threatening nettles. The kingcups illuminated the waste land; as long as their roots were wet (and they were, the reed-bed was sodden), they did not mind fighting for their right to exist. There was clump after clump of them, alight with buttercup-yellow flowers, the stems branching with three or four flowers to each branch so that there were sometimes seven flowers on a single stem; it would have been wearisome to count the total number of flowers to a clump: my guess was about a hundred. Beside this kingly splendour, the yellow dandelions flowering furiously along the dry paths looked dull, ragged. You would think the dandelion a primitive flower and the kingcup a refined one; in fact, it is the other way round; the kingcup is thought to be one of the most primitive flowers there is, with its five sepals, no petals, no nectaries, loosely bunched stamens and, in the centre, a ring of nectar-secreting carpels. But there is intrinsic beauty in the cup-shaped flower commemorated in the botanical name (for *Caltha* comes from the Greek *kalathos*, a goblet), and intrinsic pleasure in the roundness of the buds (commemorated in many of the common names – not only Marybuds but Mary-blobs).

Beside the spectacular beauty of the flowers on the fen, our little pond-bound specimen of *C. palustris*, which came from a local goldfish farm, is still a nonentity. But if I had a natural pond, the first things I would plant along its edges would be kingcups. I would seek to reproduce the natural, wild picture – the spreading clumps of dazzling yellow rising over glossy green.

BUTTERCUPS

Ranunculus

May is the month of apple-blossom, tulips and wallflowers, lilac and laburnum, wisteria, cow-parsley, buttercups and daisies. On a sunny morning when the garden was full of the scent of blossoming trees, we set out to poison the weeds on the lawn – *Ranunculus repens,* the creeping buttercup. We had a can with a sprinkler bar, and our prey was everywhere underfoot, spreading in thick mats of palmate leaves, prepared to take over square metres of space. The leaves were coarse versions of anemone leaves, the flowers small versions of *Caltha palustris*, the kingcup. But we hated them for their health, so much so that we would poison them, even though in general we considered ourselves to be against poisons in the gardens. And we knew, as we sallied forth, that we would not win: the leaves would look surly for a bit but would revive next year.

I have picked one or two flowering stems and brought them indoors. At first, they appear to be quintessential members of their family, but then I discover five colourless little sepals behind each shining yellow flower, and have to admit that these five brilliant yellow segments may be petals. I used to be annoyed when I was told that what seemed to be petals were really sepals; now I am quite put out when I find a member of the family Ranunculaceae with petals, especially when they belong to the type genus, the one that gives the whole family its common name. In other respects, this flower recalls the general pattern of the aconite and hellebore: in its centre is a domed receptacle below which fans out a bouquet of free stamens; if you look very carefully indeed you may even discern tiny nectaries between stamens and petals. The soft yellow pollen falls on to the glistening yellow petals, whose light-reflecting surface attracts the insects. These lacquered petals attract me, too. A fat old lady gave me a bunch of buttercups when I was small, and taught me to love them, and that love still runs underneath the hatred that prompts me to water them with weedkiller.

There is another weed in the genus *Ranunculus*: the lesser celandine, *Ranunculus ficaria*. It spreads by little bulbils and by seed. People write persecuted letters to gardening journals asking how to get rid of it; they are advised to dig out each plant with its surrounding soil and put it in the dustbin. Others allow it all the room it wants: the garden designer John Codrington even stuck a label saying '*Ranunculus ficaria*' beside it in his garden to keep weed-spotters quiet. Wordsworth wrote two poems in its honour. It sits very close to the ground when it suddenly appears in spring; its tiny, heart-shaped leaves carpet the bare earth so prettily that wherever it sits that corner will look nicer with the celandine than without it. Its yellow flowers shine as brightly as a buttercup's and then, just before it might start to get on the gardener's nerves, it quietly disappears. I have acquired a very pretty white one and a rather *outré* one with purple leaves. But my favourite celandine placed itself on the edge of the path that leads up to my front door, and I take on the role of Wordsworth's thrifty cottager who:

> Joys to see thee near her home,
> Spring is coming. Thou are come.

ABOVE: The celandine opens wide its shining flowers in spring.

RIGHT: *Ranunculus asiaticus* has rounded, many-layered flowers with black centres like poppy anemones.

The word 'ranunculus', however, suggests something very different from buttercup or celandine to me. To me, ranunculus is a substantial florist's-shop favourite – the turban ranunculus, *R. asiaticus*. It is well-named, a plump, folded turban of a flower, many-layered; William Robinson called it 'dressy'. It can be white, or buttercup-yellow, or sealing-wax red or the pink of raspberry sorbet. In my Australian childhood everybody loved it and took it for granted. It grew a foot tall and its leaves were typically dissected buttercup leaves. It was easy to grow in that warm climate, though it was commonly treated as a bedding plant and was used to give body to mixed vases of flowers – just as the Dutch flower paintings of the seventeenth century and the French flower paintings of the eighteenth century used it. Only in England, it seemed not to exist. This was not because it was half-hardy, difficult to grow. It was simply because it went out of fashion. The Asian ranunculus is no more difficult than the dahlia. It grows from a tuber which needs to be planted in early spring and lifted and dried and stored when flowering is over, ready to be planted out again next year. In the eighteenth century, this process was a commonplace. The best-selling gardening handbook of the day, *Everyman his own Gardener* by John Abercrombie (1767), assumes that Everyman will have a ranunculus bed in his garden, three or four feet wide (a metre or so), in which he should plant six or eight rows of ranunculus roots; alternatively he can arrange small patches of ranunculus among other flowers – 'in a circle of about six inches diameter you may plant four or five roots,' he writes, 'that is, one in the middle, and the rest round the extreme parts of the circle.' During the century more and more varieties of ranunculus appeared; some were Persian, some were Turkish, some Scottish, some Dutch, until finally the hybrid varieties were numbered, not in scores, but in hundreds. And then it all became too much; gardeners were satiated with ranunculuses; the pendulum of fashion swung.

But I wanted to see if I could grow them. I found them listed in Kelways' catalogue and ordered ten of 'Kelways' new improved strain of Super

Hybrids'. They duly arrived through the post and looking unpromising enough: small, dry, straw-coloured, with pronounced claws. The instructions said you should soak them for a few hours and plant them, claws downwards, 5cm/2in deep. Then you would be rewarded with 'a perfect riot of gorgeous beauty in May–June'.

While I was waiting I came upon the riot of gorgeous beauty at a flower show. Ranunculus were in fashion again. They were bigger and more cheerfully clashing than any I had ever seen; they were called 'Bloomingdale' variety, and were like small, voluptuous peonies, or absurd artificial flowers made out of silk. A week or two after seeing them thus, I came upon them again, in a glamorous display in the garden department of a supermarket. People were being drawn towards them as insects are drawn to the glistening petals of the buttercup. Slightly larger than life, they looked more suitable to a town courtyard than to a cottage garden, but I found myself with six of them and planted them in two flower pots, three to each. One could admire these through the cottage windows, and my grown-up children saw them thus and pointed their fingers at my lapse in taste. 'The garden is lovely,' they cried, '*except* for those!' I had evidently brought them up to approve only soft colours, subtle blendings, so this mixing of pink and yellow with scarlet appalled them, while for some perverse reason it pleased me, though the yellow shouted down the lingering daffodils and made the red japonica on the wall look coral-pink.

The 'Bloomingdales', well-watered, flowered without cease. Each globular flower was made up of innumerable shell-shaped petals, held inside coloured sepals and arranged in concentric circles and curving slightly inwards towards the centre. Full-blown, they opened out and reflexed to show a central eye of green or black, and now they began to look like poppy anemones. You would think their basal leaves belonged to an outsize buttercup, but the leaves on their stems were finely cut. As one flower finally dropped its petals, two fat buds were waiting on the same stem to take its place. The show seemed set for a long run.

Meanwhile, very quietly at the bottom of the garden, far away from the piercing brightness of the flower pots, another sort of ranunculus opened in May. Its flower was distinguished not by largeness, but by smallness, not by vibrant colour but by whiteness, though when I came upon its opening buds I was specially charmed to find that they were not pure white, but deliciously cream. The pure white came later, when the massed buds opened to make tiny flowers growing in threes (or occasionally fours or twos) on stiff, branching stems, each smaller than my thumbnail, round as a pearl, and so neat and chaste that its common name is Fair Maids of Kent, or Fair Maids of France. Its botanical name is *Ranunculus aconitifolius* 'Flore Pleno', though its leaves are less like an aconite's than, simply, a buttercup's in a handsome, dark green version toothed round the edges. It is the darkness of the leaf behind the pure white of the flower that gives it character.

I had read about it but never seen it. All the writers agreed that it is a treasure, easy to grow, that ever since the sixteenth century it had been a cottage-garden standby but now, for some unaccountable reason, it was out of favour. I had ordered it along with *Adonis amurensis* in a spirit of curiosity last autumn, and when its leaves began to appear in March I sensed that it would ultimately become a strong clump, as the books had promised; all it needed was cool moisture, in shade or even sun. As May developed into record-breaking dryness, I carried the occasional watering-can down the garden to encourage it; it never looked less than healthy, and its stems branched stiffly outwards like the caltha's – plum-coloured above and pale green beneath.

The tiny double flowers were made up of layer upon overlapping layer of rounded petals. 'Perhaps it is simply a miniature version of the Asiatic ranunculus,' I thought, and picked a small white flower-spray to see what the magnifying glass might reveal. But on close inspection it was not the same; it was tidier, tighter than the 'turban', its petals curved away from the centre, not towards it. It made the 'Bloomingdale' ranunculus look positively loose and frilly.

One day, perhaps, people who plant white gardens will see how useful it would be to them, under a white-flowering *Viburnum plicatum* for instance, as a spreading ground-cover dotted all over with a lavish sprinkling of tiny snow-white rosettes.

I shall have to mark my Fair Maids of France with a stake and mulch it well before, like the adonis, it becomes dormant. Next year I believe it will reappear a little bigger, and slowly, over the years to come (again like the adonis) it will form a larger and larger clump until it is 60m/2ft tall and wider than its height, its fleshy roots reaching outwards as well as downwards from its crown, a useful ground-cover from March to midsummer between white rugosa roses with a pale yellow tree lupin beyond.

Now I have a double yellow buttercup – not a Fair Maid but a Bachelor's Button. It is the double form of the meadow crowsfoot, *R. acris* 'Flore Pleno'; it is a brave border flower, just as tall and floriferous as Fair Maids of France, just as easy and well-behaved. It does not invade; its long wiry flower-stems rise from a mat which needs dividing from time to time. I picture it serenely blowing above the uninvited common buttercups in the wood. Its tiny flowers have green centres, absurdly reminiscent of the double yellow, green-centred adonis flower (*A. amurensis* 'Flore Pleno') and it is also uncannily like a miniature version of the double marsh marigold (*Caltha palustris* 'Plena'). At this vacant moment between spring and summer it cheers me up.

For May has ended by turning cold on us, windy and cloudy and dry. Waiting for rain, we have put away our weed-killer, for the poison needs to be applied to damp ground. The buttercups have gone on spreading and flowering, filling the edges of the waiting garden. I find myself wishing that I had still more of them, rarer ones like *R. gramineus* – lemon yellow with glaucous grassy leaves – or the charming *R. rupestris*, whose tall stems are crowned with flowers 5cm/2in across. But I have only to go through the hedge and round the pond to come upon a flower which looks like the most beautiful buttercup of them all.

SUMMER

GLOBE FLOWERS

Trollius

It is not a buttercup. It is a trollius. But I picked a buttercup to hold beside it and the yellow was identical, the close relationship clear. Only the stately trollius was infinitely more beautiful – the apotheosis of the common buttercup, its ascent into heaven.

Though not a species of ranunculus it is, of course, a member of the family Ranunculaceae, as one glance at its leaves, as well as its flowers, makes plain. But it rises straight; the verticality of its stem contrasts with the sideways-slanting habit of buttercups and marsh marigolds. Each tall stem is crowned with a large single flower, but much lower down each stem another bud is forming, and lower down still, another. Perhaps because of the dryness of the season and because I failed to cosset it with my watering can, it had only one flowering stem this year, though last year it had several. Meanwhile, the dryness of all the soil round the pond had encouraged a long crack in the concrete – on the opposite side from the trollius, so the

Trollius europaeus, Britain's native globe flower.

seeping damp did not do it any good. The water level was falling, the marsh marigolds only just had their toes in the wet; it should have been a sorry sight – one of those gardening set-backs that make the whole game seem scarcely worth the labour – but the single flower refused to look pathetic and, instead, adorned the sinking pond.

Its secret is simple: it is perfectly round – the globe flower. Roundness is a recurrent theme in the family. The winter aconite is round in all but mild weather; Asian ranunculuses are globular, Fair Maids of France are spherical pearl buttons, the buds of marsh marigolds are round as marbles. But the globe flower is round even when in full flower. Lyte described it, in the sixteenth century, as:

> a strange kind of ranunculus, the flower growing uppermost of the stalke, yellow like unto gold and fashioned lyke unto other ranunculus, but bigger and not whole open, but abiding half-shut.

The globe is composed of up to ten yellow sepals, cupped to contain the same number of incurved petals – or honey-leaves, as some botanists prefer to call them; these are thin and pointed, smaller than the sepals, not much larger than the stamens. Each segment overlaps the last, as in a hand of playing cards, and the whole arrangement is a spiral round a central convex base. But after two or three weeks the sepals fall, leaving the wispy honey-leaves exposed and vulnerable. The globular shape now falters; the honey-leaves curl, loosen and fall too. Christopher Lloyd has said that the brief flowering of the trollius disqualifies it for a place in the herbaceous border and recommends fitting it in between shrubs, for it enjoys half-shade as well as sun. But for anyone with a pond, the pond's edge is the place for it. It will *look* right, and the attentions of the watering can will make it *feel* right. After a year or two, it can safely be divided; it grows from fibrous roots, so in autumn or, better still, at the start of new growth in spring, it can be dug up, coaxed into two halves, and two plants can be quickly returned to the pond's edge where only one grew before.

It is early summer's answer to the marsh marigold of spring. It, too, needs moisture, though not mud; it is a native of damp meadows. Dorothy Wordsworth found it growing beside the lake at Grasmere, dug some of it up, along with wild strawberries, and took it home to plant in the Dove Cottage garden. She called it 'Lockety Goldings'. Gerard called it 'Lockron gowlons'. Confusingly its common names of May-blobs and Maybuds also echo some of the caltha's. *Trollius* derives from the German *Trolle blume*, which may come from the German *trol*, a globe, or may mean 'goblin flower', for there could be magic in its golden circle.

It would be life-enhancing to come upon it growing wild beside a lake, or in an alpine meadow, or on a Welsh mountainside. A Victorian doctor, Forbes Watson, described the experience emotionally when he was dying:

> Wherever we meet it, it commands our instant homage. Amidst the blaze of gaudy flowers, none looks of a descent more manifestly noble . . . The jolly buttercups and field flowers appear like country folk; it stands among them all conspicuous like a king.

This recalls E.A. Bowles and his *Caltha polypetala* – the 'emperor'. To these Victorian gentlemen, great naturalists and amateur botanists, the cup-shaped flowers of caltha and trollius were not so much magic as royal.

It is the species *Trollius europaeus* that grows wild in Britain. I would ask for nothing better in my garden; it is clear, pale lemon yellow. But most garden varieties are called *T.* × *cultorum* and are hybrids between *T. europaeus* and *T. asiaticus* or *T. chinensis*. They fall, simply, into two groups: the yellow and the orange. The European parent supplies the yellow, the Asiatic supplies the orange. The varietal names describe the colours. On the one hand are 'Canary Bird', 'Bressingham Sunshine', 'Lemon Queen'; on the other are 'Orange Princess', 'Fireglobe' and 'Salamander'. The orange trollius has such mesmerizing luminosity it would make a good buttonhole for a pedestrian on a busy country road; sometimes the colour of tinned apricots, sometimes of marigolds, sometimes a sort of burnt sugar, but always with an electric charge. It needs careful placing among cool, varied greens. But the yellow are easy; I have seen them planted with turquoise meconopsis, and with deep blue bearded irises, and read of them mixed with 'foaming stands of smilacina' (Stephen Lacey, *The Startling Jungle*). But they do not need to be mixed with other flowers; they ask nothing better than a background of leaves, silver and gold as well as green. Mine happen to have the spinach-green of *Crambe cordifolia*, the creamy-yellow of the striped grass *Glyceria maxima* 'Variegata' and the shining silver of the Scotch thistle, *Onopordum acanthium* behind them. Forbes Watson, in his dying reverie, said the flowers made him think of 'some strange metal in which gold and silver are combined' – in which case a gold and silver background picks out the component elements of the flower.

The closer the colour comes towards the silver, the better I am pleased. There is a lovely pale cultivar called *T.* × *cultorum* 'Alabaster', though it is said to be not as robust a grower as some others. I saw a beauty, pale as Continental butter, in Alan Bloom's garden at Bressingham, but its label only said 'Number 2'. I hope this means they are continuing to hybridize the trollius at Bressingham, as they are elsewhere. Varieties come and go, the list grows longer.

There are twenty species of trollius altogether, but I have only seen two others. *T. chinensis* to my eye is indistinguishable from *T. europaeus*, though I choose to call its colour 'Chinese yellow' instead of 'buttercup yellow' (E.A. Bowles was excited by it in its orange version); *T. pumilus* 'Wargrave' is a miniature, only 15cm/6in tall, with clear yellow flowers which are more open than globular; it is not specially like other trolliuses, but it is immediately recognizable as a member of the family.

By now these family resemblances overwhelm me, and I begin to wonder why I ever thought the classification of the Ranunculaceae puzzling. More often than not, it is the leaves I recognize first: much-divided, toothed and lobed. Their botanical name is 'pedate', which means 'having divisions like the claws of a bird's foot'. No wonder the trollius used to be called 'globe crowfoot': its leaf is divided into three, and each division is divided again. The central bird's claw is elaborately cut into three along the top, while the two side claws are divided so deeply again that the cut goes down almost to the base, giving the impression that the whole leaf is divided, not into three, but five. When fully grown, the divisions of the leaf spread wide like the fingers of a hand at full stretch so that the effect is not so much leaf-shaped as circular, recalling the deep-cut circular leaves of the winter aconite.

The evidence of the leaves, added to the simplicity of rounded flowers with coloured sepals and a sunburst of stamens, makes the game seem easy to the point of monotony. But I am only halfway through the gardening year. It is June; in June comes the columbine, named after a ring of doves, and simplicity of shape flies away.

COLUMBINES

Aquilegia

Take a simple, star-shaped flower. Give it five coloured, pointed sepals opening flat round numerous stamens, just like certain clematis. Now give it five nectaries, arranged inside the sepals and straddling the junction of one with the next. The flower is now structured like the hellebore or winter aconite. But now set to work on the nectaries; fold them into horns or cone shapes, slot each one through between two sepals and roll it into a spur. You will have five slender spurs, full of nectar, at the back of the flower; you may finish each one off with a little knob. In the front, mould the cones into deep, dimpled cups, inviting bees to come and sip – and these will be called 'petals'. The result will be the most elegant and lyrical variation on a simple theme, as difficult at first to recognize as it is to hear the original theme in many a free-ranging musical variation. But once the theme is rediscovered within the variation, there is extra delight in it. The columbine metamorphoses a simple flower shape into an intricate one, and throughout its history people have compared it, not to other flowers, but to birds.

'Columbine' derives from *columba*, a dove. The *OED* records that it is thought to resemble five pigeons clustered together in a ring. If you turn the flower upside down, you can see their heads, and understand why it used to be called 'Doves-round-a-dish'. Aquilegia recalls *aquila*, the Latin for 'eagle'. An eagle's claw was seen in the curved tips of the spurs. The wide spread of the sepals was the bird's spread wings. If you pull off all the sepals but two, and all the spurred petals but one, you will clearly have a bird in your hand.

But the flower plays a trick on us. The long spurs at the back borrow colour from the sepals as if the paint had run. The result is that, unless you look very carefully, you see the spurs as belonging to the sepals, and not as extensions of the petals at all. This is the final complexity in the aquilegia's variation on a simple theme.

The flower that I have imagined being made is *Aquilegia vulgaris*, the European native that led Linnaeus to the discovery that these long, tubular petals are, in fact, nectaries. (To less scientific people, the spurs had long been seen as emblems

ABOVE: A pure white columbine with fly-away spurs.

RIGHT: *Aquilegia* 'McKana's hybrid': the spreading sepals of these columbines are raspberry-red, the dimpled petals are primrose-yellow, extending at the back of the flower into elegant spurs of the same colour as the sepals.

not of sweetness but bitterness: they represented a cuckhold's horns.) *A. vulgaris* is the original 'columbine', and comes in blue, pink, purple or white. The white is specially treasured in a form called 'Munstead White' (properly *A. vulgaris* 'Nivea') with frosted bright green leaves. But a hundred years ago its supremacy was challenged; brilliant aquilegias arrived from America called 'Mrs Scott-Elliot's strain' or the 'McKana Giants', both bred from the Mexican *A. longissima* whose spurs extend at the back to a tapering point, doing away with a pigeon's-head knob. These are the long-spurred hybrids of the seed packets, and the lovely flowers of most people's gardens.

They are nearly always two-coloured, with the paler colour for the petals, the darker for sepals and spurs. Bright colours arrived with the American strains, particularly red with gold. But in the long-spurred F_1 hybrids on trial at Wisley lately, there were also combinations of white with blue, white with pink, cream with white. The inclination of these aquilegias is towards the pale; bumble-bees get to work on the bold Americans and the resultant nameless seedlings fade into sunset colours: first apricot and daffodil yellow, then cowslip touched with coral-pink. Pale yellows and pinks predominate, which is partly why I particularly treasure the blues.

Most of my aquilegias come from Suttons' long-spurred hybrid seed packets and my favourite has sepals and spurs of smoky lavender, and primrose petals. But I also have a handsome plant called 'Crimson Star', where the sepals are not crimson at all, but soft raspberry, and the petals are white.

A year ago at Chelsea I saw a whole stand made up of aquilegias called 'Dragonfly'. It was a ravishing mixture with lots of white among the pink and lilac. I came away with a very small brown envelope of seed, and sowed it in the vegetable garden. Only three plants survived the vicissitudes of our seed-bed. I moved them to their flowering positions and told myself that, since aquilegias are poisonous, the rabbits would know better than to eat them. I was wrong. I began to find whole sprays of ferny leaves nipped off. I put wire netting round them, which completely obscured the leafy grey-green beauty of the clumps, and waited to see which colours my dragonflies were.

Aquilegia stems come in variable lengths. One of these seedlings made flower-stalks that barely cleared the rabbit wire: then the flower panicle began to open; the flowers were entirely violet, petals and sepals alike, and there were no spurs to be seen. Aquilegias without spurs are now sometimes called 'semiaquilegias'. This semiaquilegia spread wide its violet sepals like wings prepared for flight; I suspect it of being a form of *A. vulgaris* called *clematiflora*, rightly underlining the family likeness between certain clematis and columbines. (Its newer name of *A.v. stellata* acknowledges its starry shape but loses touch with this important comparison with clematis.)

The second seedling did not produce any flower-stems at all. The third grew ridiculously tall on its wiry stem, almost 60cm/2ft. Daily I waited for its slender pointed buds (like clematis buds) to open, and when at last they did they were not dragonflies as I remembered them, but a double form of granny's bonnet. I did not know whether to be pleased or angry: granny's bonnets are nowadays once more high fashion.

The sort of bonnet granny is supposed to have worn has a pale, much-ruched and pleated frill inside the brim, framing the face. The brim is the columbine's sepals; petals compose the frill. The sepals fall first, before the petals, and when they go you can observe the entire shape of the petalled frill from front edge to spur, all fluted as if fresh from a goffering-iron.

Granny's bonnets come usually in purple, blue or white. But mine were pearl-pink with a white frill. The slender wiry stems branched and branched again, each branch carrying three to six flowers which pointed this way and that, like hats displayed on hat-stands in a millinery department, all at different heights and facing in different directions. The flowers had great charm, and a touch of the absurd, so small and muddled on the tops of their long, strong stems. But for elegance of line and fly-away gaiety of spirit I stick to the long-spurred hybrids, though the pendulum of gardening fashion may be swinging away from them.

It is swinging towards *A. vulgaris* 'Nora Barlow', a new name for an old species – the 'rose columbine'. Her concentric, overlapping segments, pointed and crisp, make her look more like a hedgehog than a bird – *en brosse*, with each raspberry-pink spine tipped in cream. The surrounding sepals are the creamy-green of a forced winter lettuce, reversing the usual colour scheme where the sepals are darker than the petals. There are no spurs to be seen at the back of the flower; perhaps it should be called a 'semi-aquilegia' too. Some people think it is like an astrantia, a spiky pincushion. Its combination of oddity, sophisticated colour and exuberance (for it is tall, floriferous and healthy) makes me indulgent towards the single specimen growing in the garden.

I also have a single specimen of *A. glandulosa*, whose spurs are hooked like walking-stick handles, whose petals are lined with dove-grey and whose sepals are lavender. But altogether there are a hundred species of aquilegia. Some have very small flowers on top of very long stalks, some have quite large flowers on very short stalks. There are some real dwarfs in the genus, cute rather than graceful. There are some wispy, spidery ones more like insects than birds; there are exotic ones from distant lands; the little flowers of the delicate *A. formosa* are tomato-red all over. There are also very dark ones with names like 'Black Beauty', perhaps explaining John Clare's line:

The Columbine, stone-blue or deep night-brown . . .

But I am not making a botanical collection. I am simply filling all the space in a certain border with columbines.

A whole bed of mixed aquilegias is a tempting thought for a large garden in June. There would be nodding heads, held high above the foliage, in violet and lavender and apricot and peach and primrose and raspberry, lightened with white. But better still is Vita Sackville-West's vision of them in a courtyard with farm buildings on three sides; they would come up everywhere, along walls, from cracks between paving stones, half-wild; she saw them, indulgently, as little more than weeds.

To mix pastel colours is more rewarding than to arrange them in separate blocks. Worst of all is to plant out a single variety in a straight row. I have seen this done with a red and gold cultivar in a neighbour's garden. The plants are spaced out a yard apart along an otherwise empty, well-weeded border, and each one is tied to a bamboo stake. It is like a death sentence.

Tall though the aquilegia is, if it is in a sheltered position it does not need staking. That is one of its beauties. But if a strong wind threatens its poise, a discreet twiggy stick or two, pushed down among the leaves, should rescue the leaning stems. My aquilegias fill spaces between hostas, and the shining new season's leaves of hellebores, with foxgloves rising behind them. Other people suggest other companions for them: bearded irises, oriental poppies, catmint, lupins – all the lovely border flowers of June. They are easy mixers; they can fit into small spaces; they are too discreet to clash with neighbours, and too distinct and eye-catching ever to be quite lost in the crowd.

Yet, in the end, they have no need of neighbours. The symmetry of the flowers asks for undivided attention and close-up viewing. Dorothy Wordsworth found one growing in the shelter of trees, between rocks, in the lakeland landscape where she had found her globe flower a year before. She decided that it was, by nature, a solitary plant – 'a graceful creature, a female seeking retirement, and growing freest and most graceful where it is most alone.' She observed that it prospered in shade.

She was right. *Aquilegia vulgaris* is a native of calcareous woodland, so shade, or speckled sunshine, is its natural element. It likes the same conditions as many of its relations: a moist, well-drained soil. Its woody rootstock goes deep, a tap-root which foils any plans to move it. But there is no need to move it, for its life will probably be short. This is the sadness of aquilegias: they are ephemeral. Some people think they start to deteriorate after their first year's flowering; some even treat them as biennials, refurbishing with new seedlings sown in early summer of the previous year. I sow fresh seed from time to time, but I never dream of throwing my old friends out; some of them disappear quickly but some, like my smoky blue, hang on for years.

They make up for the brevity of their life-span by the length of their flowering season. It is said, if you choose your species cunningly, starting with the alpine ones, that you can have aquilegias in bloom from April until July. But a single generous plant, whose first buds open in mid-May, can go on producing flowers with marvellous prodigality for eight successive weeks. And all the time, as the individual flowers fade, you must dead-head them. If you allow the seed-heads their way, the resultant self-seeded plants will be wishy-washy, altogether inferior to the lovely colours truly pictured on the front of the seed packet. Aquilegias are well-known for promiscuity; there is no genus whose different species so enthusiastically interbreed. It would be sad to lose 'Crimson Star' and *A. glandulosa* and end up with nothing but palest yellow tipped with coral.

Parkinson almost despaired over the way the beautiful columbines in his garden died out, leaving him with the duds. Yet he could not relinquish them. 'Columbines', he wrote, 'are flowers of that respect, as that no garden would willingly be without them, that could tell how to have them.'

We love them for the shapes and colours of their flowers. Their blue-green leaves are a bonus, loved for shape and colour too. They are not always glaucous; they can be fresh green or, at worst, speckled yellow (as in 'Granny's Yellow'). They are usually divided into three and three again ('biternate' is the correct word for this formation); each division is like a separate leaf so you end up with a delicate spread of rounded and scalloped leaves, each with just enough concavity to hold a raindrop in the middle. Sometimes you find a lobe that is divided not into three, but four – a bit like discovering a four-leafed clover. The beauty of the thing defies analysis. I have to force myself to pick one and try to grasp the fact that the leaves belong together in a pattern formed by a compound leaf. The aquilegia's leaf is as much a fanciful variation on the typical Ranunculaceae leaf as the flower is a variation on the typical Ranunculaceae flower.

'There's fennel for you, and columbines,' said mad Ophelia to bad Claudius. The fennel stood for bitterness, the columbines for ingratitude. Like most desirable things, the sweet columbine brings opposites together; it is both the Eagle and the Dove.

Aquilegia 'Nora Barlow' has concentric, overlapping segments of pink and creamy-green, like a spiky pincushion.

MEADOW RUE

Thalictrum

There was an untidy arrangement in the vase on my window-sill: it was composed of two leaf-stems, one pointing one way and one the other. The stem pointing left was subdivided into threes and threes and threes again, until it finally produced trios of small rounded leaflets, roughly heart-shaped except that they were finished off with three rounded scallops at the top. The whole many-branched and pinnate affair was both stiff and soft, green tinged with blue, as decorative as maidenhair fern. In comparison, the leaf-stem facing right was rather floppy and untidy, a brighter shade of green, with larger, less completely subdivided leaves; the divisions were lobes rather than leaflets. Yet the left-hand leaf was called after the right: it was *Thalictrum aquilegiifolium*, and I had picked it to compare with its namesake, the aquilegia leaf on the right. If this had been a beauty contest rather than a botanical examination, the thalictrum leaf would have been a winner, lovely though the aquilegia was.

But thalictrums have always been valued for their leaves. When we came to our Norfolk cottage nearly thirty years ago, there seemed to be no plantings remaining in the garden at all. Brambles, nettles, elders, thistles and rough grasses ruled. But after a year or two some pretty, ferny, fresh leaves emerged; they were the cottagers' favourite, the poor man's maidenhair fern, and they proved irrepressible. It was years before I discovered, from Margery Fish's *Cottage Garden Flowers*, that the ferny favourite was called *Thalictrum adiantifolium* (syn. *T. minus adiantifolium*). Cottagers, she explained, liked to use its leaves as a background for buttonholes and posies, but when she said that it had a network of creeping yellow roots that were impossible to eradicate, I knew she was talking about my plant.

It spread right along the cottage wall. I pulled it all out (as I thought) and because it was pretty and old-fashioned I replanted a small piece of it in an empty bed near the pond which was then in need of furnishing; sentimentally I shrank from doing away altogether with the survivor of a buried past. So now I have *two* beds of *Thalictrum adiantifolium* – one by the cottage wall, one by the pond. I have soused it with glyphosate, but after pretending to be demolished, it has resurrected itself. I shall never get rid of it, though it is busy trying to get rid of anything that grows in its path. It is a mixture of trembling frailty and tenacious strength. The flowers are mere wisps: tiny clusters of lime-green stamens at the branching tops of tall, wiry, polished stems, green brushed with mahogany. There are basal leaves and wide-spaced sprays of leaves right up the stems, all dainty and ferny. But I do not pick them (I wrench and crush them). There are 130 species of thalictrum, many of them described by Christopher Lloyd as 'rubbish'; this belongs with the rubbish. But the three species of thalictrum that I have chosen to grow belong with the riches of the garden.

Not that the flowers are rich; they seem, on first acquaintance, like B-class flowers – shapeless, mauve or green, composed of soft amorphous fuzz; they look born to be foils, subtly enhancing the beauty of A-class flowers. But the allure of the thalictrum comes, not from the flower, but from the whole plant, its height, its deportment, its long season – and its foliage.

The leaf in my vase belonged to *T. aquilegiifolium* 'Thundercloud'. The name is not a hit, for the flower of this variety is bright mauve, leaning more towards the pink than the thundery blue. The colour is spread evenly over a dense panicle not unlike a small version of a rhubarb flower, which may be why in Gerard's day thalictrums were called 'bastard rhubarbe'. When I first planted it, I was tempted to ignore the flower and dwell upon the shapely, glaucous leaves which seemed to get bigger and lovelier as they mounted the 90cm/3ft stem. But when at last I looked closely into a tufted panicle, I discovered that the 'thundercloud' was made up of a series of small explosions, bunches of stamens released from their sepals and bursting into tiny purple spheres round their respective centres. There were dozens of them making up the panicle; the ones at the top were small round balls, still held in place by four parma-violet sepals, waiting to explode. It was a revelation – a further variation on the Ranunculaceae theme; here was a flower with no petals, four coloured sepals and a multitude of prominent stamens attached to the base of the carpels. The stamens made the fuzzy flower; the sepals fell as soon as they opened, and the colour came from the stamens' long filaments, broadening at the top where they were finished off by little dabs of creamy-green, the anthers. Each tiny flower was vibrant and complete.

I first planted *Thalictrum aquilegiifolium* under a Mount Etna broom. Then I read that it looks good planted among coppery-purple shrubs. So I moved it to a place between the only two purple-leaved plants in my garden: *Cotinus coggygria* 'Notcutt's Variety' and *Berberis* 'Atropurpurea Nana'. It submitted to the move without complaint, but to my eye the purple leaves of its neighbours drained its beautiful glaucous leaves of life and accentuated the uncompromising fierceness of its flower's mauve. It absorbs rather than reflects light, so it needs bright leaves or pale leaves around it: sharp and shaded euphorbias, silver senecios, striped grasses or irises. I picked a stem and wandered with it round the garden, following the practice recommended by Vita Sackville-West, trying it first here, then there. Finally, I thought it might enjoy the company of its cousins; I stuck it in among the long-spurred columbines and it was instantly embraced and tamed by soft pastels, so much at home that its separate identity was in danger of disappearing. I understand why Parkinson called it the 'tufted columbine'.

There is a white version, *T. aquilegiifolium album*, on offer, and the type, *T. aquilegiifolium* itself, a subtler violet than 'Thundercloud', as well as being a taller, more stately plant, though sharing its firm stance and early flowering: it starts to bloom with the first long-spurred hybrid aquilegias in mid-May, and like them is still there at the end of June. In July, its seed-heads are stiff cobwebs, too light and lacy to cut off. But even better thalictrums are on the way.

The famous, favoured thalictrum does not flower until July or August. It is *T. delavayi* (often called *T. dipterocarpum*), noble in height, infinitely delicate in detailed silhouette. It has the sort of flower that can confirm a child's belief in fairies. The individual florets are minute, poised on wiry stems sufficiently far from each other to make the whole pyramidal flower-head a pattern in space, each tiny floret seeming suspended in air. These florets are made up of four or five coloured sepals and a burst of coloured filaments, just like *T. aquilegiifolium*, but because the flowers do not touch each other, the mauve never looks solid or sickly, and fluffiness is avoided.

The plant was discovered in the nineteenth

ABOVE: *Thalictrum aquilegiifolium* grows tall and strong, with fluffy mauve flowers.

RIGHT: *Thalictrum speciosissimum* has the strength and grace of a meadow wild flower.

century by the Abbé Delavay; its flowers have lilac sepals and white or yellow stamens; it is 1.5m/5ft tall, yet with a disproportionately modest rootstock. It has the temperamental nature that befits a prima donna; each long stem needs to be tied to a short stake which will be hidden by the branching leaves; so long as its base is firm, the stem is then able to soar upwards and burst into its wide-spreading display, 60cm/2ft across. But it will not perform well without pampering; it wants rich food and plenty to drink; light, moist, well-drained soil and a touch of sun. Unless these things are supplied it will shrink in successive years and finally disappear.

I grow *T. delavayi* 'Hewitt's Double', which is a little shorter than the type and no less temperamental, but even more beautiful because its filaments are not white but a pale sweet heliotrope. It takes its time about reappearing in spring, so that unless you have marked it with a stake you wonder precisely where it used to be last year. When it comes through in May its little ferny leaves are feebly inconspicuous. Very gradually they grow, and their wiry stems are chocolate-coloured; then flower-stems develop, very thin, very hard, jointed at intervals with leaves growing from the joints. If it is happy (which mine tend not to be) its leaves metamorphose from grey-green to pea-green and fan out expansively below the hundreds of closed globes and open powder-puffs. One of the charms of the flower-head is its mixture of closed buds and open, slightly frilly flowers. There is an exquisite white version with fresh green leaves, white stamens, and pearly sepals that are always threatening to revert to a ghostly mauve.

T. delavayi lasts beautifully in water, and in a mixed vase it contrasts with more solid, round-faced flowers just as gypsophila might, though the comparison seems to vulgarize the thalictrum; it is no mere in-filler. If it can be picked with a very long stem its beauty will dominate a vase of flowers placed on the hearth in front of a black grate. How best to place it in the garden is another question. Perhaps it doesn't care, so long as it is somewhere it can be admired; it likes to be alone.

There is no question of dividing it. It does not spread enough to make a divisible clump. It is a plant to buy, love, pick, watch anxiously if it fades out, mourn, and then replace expensively from a plant centre or, enterprisingly, from seed. But when it is flourishing, it flowers for weeks and weeks, all through the second half of summer.

My third thalictrum is *T. speciosissimum*, 'most handsome' by name and most special to me. It is a

Spanish version of *T. flavum*, the yellow meadow rue, but *T. flavum* has green leaves and the leaves of *T. speciosissimum* are grey-blue (it is sometimes called *T. flavum glaucum*). The common name meadow rue suits it, for its leaves are the colour of *Ruta graveolens* 'Jackman's Blue', and you can imagine it growing in a meadow among tall grasses and weeds like hemlock and meadowsweet and holding its own, its fluffy heads tossing high in the wind. It is wild and towering and strong, as well as fine-cut and see-through. But I doubt if it is to be found in our meadows now. It is not even to be found very often in our gardens, because most of us choose our border perennials for their flowers, not their leaves.

This summer it was poised at the edge of the pond when a thunderstorm was gathering in the sky, and I thought it, rather than *T. aquilegiifolium*, ought to have been called 'Thundercloud'. Its leaves, matched against the sky, were deep cloud-colour, and the thin stems, or petioles, leading to the flower panicles were dark plum. The same colour was striped on the pale green stems: plum pencil-stripes all the way up the stalks. There was pale green, pale yellow, glaucous blue and plum in the plant, but from a distance everything else was submerged in the dominant glaucous blue leaves.

They stood out from the joints of the stems in tiers like arms balancing a tight-rope walker. The lower fronds drooped a little; the higher ones tilted upwards, giving the whole plant its joyous, upbeat air. Between any two joints the stems took a slight lean, to be corrected between the next two joints, so that the finished result was straight. The leaves contrasted and yet blended with everything that was growing near them: the bent-over blades of flag irises, the glossy pads of bergenia and rheum, the rough dull greys of a row of artichokes. The flowers were not yet open. Buds waited in all the leaf axils as well as at the tops of the stems – tight-packed panicles, the colour of unripe lemons or of *Alchemilla mollis*. What *Alchemilla mollis* does for its neighbours at ground level, complementing them whatever colour they may be, *Thalictrum speciosissimum* can do at a higher level among the giants of the garden border. In classic herbaceous borders, its approved role is as a foil for delphiniums.

May had been cold and dry. June was cold and wet. 'I expect my thalictrums are waiting for the delphiniums,' I thought to myself, even though I was growing them far apart. 'When the delphiniums open, the thalictrums will.' July brought hot sunshine and brilliance, a dazzling sense of high summer. The delphiniums were soaring, opening, showing off, yet *T. speciosissimum* by the pond seemed still to be only in bud. Impatiently I picked one precious stem and held it beside each of my delphiniums in turn, just to check whether it would indeed make them look more beautiful than ever. It did. They were all different blues; it paid equally graceful compliments to each. I took it inside and put it in a tall glass mug.

Then I discovered, to my chagrin, that it *was* open. The sepals had fallen; lemon-yellow anthers had appeared at the ends of long white filaments with a lingering green shadow behind them which had made me think it was still in bud. It broke into a corymb of ten flower stalks, each carrying bunches of puff-balls. The next day the puff-balls had rounded and expanded; it was making an important statement, catching the eye alone in its mug. I had been busily presenting the flower-head as insignificant compared to its leaves, though useful as a foil to grander flowers; that morning I saw its individual beauty, subtle yet sharp, cloudy yet luminous, and as fresh and fizzy as lemonade.

DELPHINIUMS

First came a delphinium of sapphire-blue, touched with amethyst along the sepal edges. It was the one surviving plant from a free-offer seed packet of 'Mr Fothergill's Pacific Hybrids' which had been stuck to the cover of a gardening magazine a few years ago. I had sown the seed sceptically in the vegetable garden and this one had escaped the rabbits and the sparrows and the mice. Now it had twelve mighty stems and the flower spikes were 45cm/18in long. It opened before June was over, and the bumble bees were delirious, burrowing into its black eye on a sunny evening after rain. Friends visiting the garden stood on the grass in front of it and took its photograph.

But that is what delphiniums are like. They demand admiration, they cannot be ignored. Yet for years I did not grow them, rejecting them on the grounds that they need staking. My husband was heard, once, wistfully wondering why we couldn't have really nice things like delphiniums in our garden. 'Because they are a nuisance,' I crisply replied. But I weakened in the end.

The mighty dark blue Pacific Hybrid stood there, all six feet of it, so straight and strong that you could not imagine it as vulnerable to wind, and unless you looked carefully, you would not see what held it up. In fact four bamboo stakes, much shorter than its stems, were hidden among them, and a string went round the whole thing only about a foot above the ground. This underpinning was done two months before and took about five minutes. The whole thing seemed too trusting to be true, and when our visitors had gone I looped a piece of raffia round the stems and canes just below the top of the foliage; the thought of losing a spike or two in a high wind was intolerable.

The second delphinium to open was exquisite cerulean blue with a white eye. This had come in a special offer of unnamed Blackmore and Langdon seedlings and its classy ancestry was obvious in the huge size of the individual florets and in the elegant poise of the pale, fat buds at the tops of the spikes. The buds were a different colour from the open flowers: misty, opalescent, as if a silver sheen were spread over palest green, pink and blue. Delphinium buds are dolphin-shaped, as the ancient greek Dioscorides is supposed to have noticed centuries ago. Gerard confirmed the likeness in the sixteenth century:

> for the floures, and especially before they be perfected, have a certaine shew and likenesse of those Dolphines, which old pictures and armes of certain antient families have expressed with a crooked and bending figure or shape, by which signe also the heavenly Dolphine is set forth.

Two more Blackmore and Langdon unnamed delphiniums were opening in the first week of July. One was dark blue, as dark as Mr Fothergill's, but with a white eye. I decided that I preferred black eyes with my dark blue flowers, and that white was too striking a contrast. But then this Blackmore and Langdon product was striking in every way: the spikes were not cylindrical, but tapered from a wide base, and the flowers were decidedly superior. Feeling like someone who catches a beautiful butterfly and sticks a pin through it, I picked a single floret from the two different dark blue delphiniums. Even minor mutilation of these mighty columns was vandalism.

But now I could not fail to see the crucial difference between them. My first was single, a saucer-shaped floret made up of five dark blue sepals arranged in a diamond shape, the top one curving back behind into a long and graceful spur, like one of the aquilegia's nectaries. My second was double; its spur was rudimentary, but inside its five sepals were ten further segments – sepals or petals, it was hard to say which, though later I saw them described as sepals. It was not a saucer, but a flat rosette, and it was as big as the flower of a *Clematis montana*. Suddenly I saw how easily delphiniums take their place in the family Ranunculaceae: I had only to study a single floret to see it. Sepals supplied the colour and shape; 'honey-petals' supplied the 'eye'. These petals were small, four in number, but there was no need of a magnifying glass to see them – their contrasting colours made them stand out; streaked indigo on the single floret, pale creamy green on the double. They were a little rakish and untidy, different shapes and sizes from each other, and the two top ones were elongated, exclamatory, like asses' ears, and fed back into the spur formed by the topmost sepals. This was a complex variation on the spurred aquilegia theme – honeyed spurs within a spur – and this was what made the bumble-bees burrow so deeply into the delphinium flower. What was missing was the great boss of stamens which had characterized each genus in the family until now; the delphinium's stamens were hidden, rather feeble-looking and floppy, within the eye.

In the long herbaceous border at Waterperry Gardens, delphiniums tower amongst yellow verbascums. Each clump is staked with pea-sticks, notched to bend inwards at the top.

Sepals and honey-petals made up the floret's beauty.

Multiply one floret by forty to eighty, and you get a delphinium spike: a tall column of apparently uniform colour, a continuous upward thrust of blue. Detail is lost in the stunning strength of the whole. Yet the colour on a single fallen sepal from my specimen floret was not simple, but as complex as a section of sky in an impressionist painting: washes of blue – bright, dark, shading to peacock, shading to purple, shading to jade green – blended into each other in a gradation so seamless that it turned the guilt felt in picking the flower into admiration and wonder.

Delphiniums are loved for their intoxicating blues. Every blue is there; all the blues the Old Masters used to clothe their madonnas, from pale to dark, from sky to sea, from sapphire to turquoise, through all the other blue flowers – cornflowers, gentians, forget-me-nots, flax, meconopsis, anchusa, morning glory. But my third unnamed Blackmore and Langdon seedling was not blue at all. It was pale dove-grey brushed with smoky pink with an ash-white eye and perversely I loved this faded delphinium best, perhaps because it grew in front of a cerise rose, and my eyes rested gratefully upon its cool racemes after the over-stimulation of the intense roses.

It was my favourite of the unnamed seedlings, but during the first weeks of this hot July a named delphinium opened in my garden and claimed my vote: 'Alice Artindale', old and famous but new to me. She was raised in Sheffield by Mr Artindale in 1935, and she was the first truly double delphinium. I now discovered that the Blackmore and Langdon seedlings, with their generous layers of sepals, are described as 'semi-double'. The florets of 'Alice Artindale' were as fully double and rounded as the flower of *Caltha palustris* 'Plena' or *Ranunculus aconitifolius* 'Flore Pleno' and for the same reason: her stamens had developed into pseudo-petals and there was no eye to the flower at all. It was no accident that breeders began to describe double delphiniums as 'ranunculus-flowered'.

The florets were not flat, but semi-globular, three-dimensional; the layers of pointed 'petals' were incurved at first like the Asiatic ranunculus, but later reflexed backwards into the shape of Fair Maids of France. Each globe had its own space on the stem and never touched its neighbour; the flower-spikes were long, tall and tapering, and there was no overcrowding. Below each central spike, graceful side spikes waited to take over. You can only describe a shade of blue by analogy with other things. Yet the blue of 'Alice Artindale' was incomparable. I have seen it described as 'cobalt', but mine was neither dark nor pale; as it matured, strokes of lavender appeared in the centre of the flowers, but in its youth it leaned neither towards purple nor towards green and I shall describe it (even though I know that there is no such thing) as 'delphinium-blue'.

Delphiniums have handsome, three-lobed dissected leaves (those of 'Alice Artindale' are rather pointy). These leaves are the first among herbaceous plants to reappear in spring. Impervious to frost, they grow steadily taller and thicker; they mean business. Aphids do not attack them, rabbits seem indifferent to them, birds are said to leave them alone, though sometimes in the past, when my flower-spikes have missed a whole series of buds below the tips of the racemes, I have wondered whether sparrows were responsible. I now believe that this is just an unfortunate quirk of certain plants in certain seasons. Miss Jekyll was angry about it: 'There is a common defect which I cannot endure,' she wrote, 'an interrupted spike, when the flowers, having filled a good bit of the spike, leave off, leaving a space of bare stem and then go on again.'

Disappointments happen with all plants. In general, delphiniums are surprisingly undisappointing. In early May, when the flower-spikes begin to form, it is time to water them. The recognition that delphiniums are moisture-loving buttercups on a grand scale has been a breakthrough for me. It is a pleasure to water them at the crucial moment in early May, and to top-dress them with a balanced fertilizer. No more need be done until the flowers are over. Fortunately there are the lateral spikes to come when the top of the central spike gives way to a series of pointed urn-shaped carpels; when these, too, are over, comes a moment of choice.

Either you dead-head them and leave the lack-lustre clump of palmate leaves to look spent for the rest of the summer, or you go for the possibility of a second flowering, and cut the stems right down to the ground, water and feed again, and wait for the indomitable plant to produce a new set of shoots. It does this with astonishing speed: barely is your back turned before new leaves appear, not as many as the first time round, and not all destined to flower, but still the pale fresh green of spring. Some experts hold that a forced second blooming puts a severe strain on the plant and shortens its life. Others pooh-pooh this warning, saying only that laterals should be removed the second time round, that dead-heading must be swift once the second flowering is over, but that then all leaves must be left to feed the crown for next year. Only when they wither should they be cut down to ground level so that water cannot collect in the hollow stems.

My impulse is to side with the optimists. Where I have cut them down, the border looks less sad; everything seems to be closing in, filling the gap where the delphiniums stood, like river water flowing into a place where a boulder has been removed. And even if I don't achieve a satisfactory second flowering, the fresh leaves furnish the border better than the old.

Healthy delphiniums left to themselves are long-lived, but they will not last for ever. The crowns thicken, become tuber-like, and finally decay. It is not easy to move or divide them. Nor is it easy to buy named varieties at garden centres; they do not lend themselves to container gardening, and look faintly absurd in pots – too tall for their age and too thin, toppling about. Specialist growers tend not to sell their named varieties in garden centres.

Specialist growers are on a quest for the exceptional – in size, shape and colour. The quest started in the nineteenth century with Lemoine in France. Long before the great firm of Blackmore and Langdon set up in Bath in 1904, James Kelway of Langport bred the first semi-doubles in England and in 1895, introduced the first ivory-white delphinium, 'Beauty of Langport'. In the 1920s, Thomas Carlisle of Loddon Gardens built up a comprehensive, international collection; in the '30s, in California, Frank Reinelt used the species *D. cardinale* to produce pink delphiniums. Meanwhile amateur growers joined in, each seeking his own ideal in colour or form. Some liked their florets to stand out from the stem on stiff pedicels; some liked lots of laterals, some did not; some were fussy about the ratio of spike length to stalk, discarding very long spikes as grotesque and disproportionate, and very short spikes as totally unacceptable; some disqualified any raceme where the lower florets fell before the top ones opened; some went for giants, others for dwarfs; small, elegantly branched, wiry plants with looser spikes of flowers, pink as well as blue, and no central raceme; these were the belladonnas and the elegant Chinese delphiniums, like 'Pink Sensation' or 'Blue Butterfly', mid-border plants which did not need staking and fitted, not the grand herbaceous border, but the ordinary suburban garden.

ABOVE: *Delphinium* 'Can Can' has frilly, semi-double, black-eyed flowers.

RIGHT: Here are the dark blue and the faded, smoky pink unnamed delphiniums growing in the author's garden. Behind them climbs the cerise *Clematis* 'Etoile Rose'.

We can join the fun; we can get seed packets from specialists, and we can raise our own. It is a lucky dip; we don't know what we may land; we can keep our favourites and discard our non-favourites, unless we want to fill up our whole gardens with delphiniums.

The sadness when they are gone remains. It comes just as the first flush of roses falls and there is a sense that the very best of the garden is over. Margery Fish thought it important to plant something nice in front of her delphinium clumps to hide the August gap: penstemons would do, or perowskia, the Russian sage, or *Phlox paniculata*, or pretty annuals. Miss Jekyll planted the everlasting pea, *Lathyrus latifolius*, behind hers, and pulled it right over the finished clumps which served as supports for the billowing, many-stemmed spread of pea-flowers and tendrils, untidy but lavish. You could perform the same trick with one of the herbaceous clematis like *C. jouiniana*.

But what to do when the delphiniums are over is less interesting than what to do when they are at the peak of their glory. For real delphinium fanciers there is no problem: you don't mix them with anything; you mass them, and stand back in wonder.

Amos Perry, one of the great Edwardian breeders, stacked 25,000 or 30,000 cut delphinium stems together at the Holland House exhibition in 1910; the result was known as the 'blue tent'. Eighty years later, growers still mass their giant spikes in rising tiers. I saw them thus at a recent flower show; it was a city of azure spires, the buds like Gothic decorations all the way up the ribs of a pinnacle, an architectural miracle but not a recipe for a convincing garden. Yet specialist growers still tend to recommend that delphiniums should be given a bed to themselves, and when they are spent, dahlias can take over.

For the majority of gardeners, who like mixing

things together and who love delphiniums but do not want to be overpowered by them, the question remains: what are their best companions? The answer is: almost anything. Blue is not difficult: it does not clash with other colours, though I do not share the dormouse's longing for:

> a wonderful view
> Of geraniums (red) and delphiniums (blue).

Yellows, rather than reds, are its ideal complement. Miss Jekyll, in her long border, mixed foreground clumps of pale blue delphiniums with yellow and white flowers, distant clumps of dark blue delphiniums with silver leaves. Delphiniums tower over the yellow and blue border at Kiftsgate; in the gardens of Clare College, Cambridge, they mix delphiniums with lemon-yellow thalictrums and the pale primrose verbascum 'Gainsborough'; in the great herbaceous border at Waterperry Gardens, there are thalictrums and verbascums again, along with the shorter racemes of *Galega officinalis* and the mother-of-pearl of *Campanula lactiflora*. One writer thinks their ideal partners are pink lupins and white lilies and fuschias; another recommends growing them among old shrub roses. The only place where they will not be happy is under trees, though they will tolerate some shade, and flower alike on sand, clay or limestone; the only time when they will not look their best is if they have no neighbours at all; delphiniums abhor a vacuum. I have seen them planted along a post-and-rail fence, yards away from one another, lonely and miserable, their beauty extinguished. The delphinium is the 'queen of the border', the 'monarch', the 'aristocrat', repeatedly described in noble terms, and deserves a retinue of supporters.

Yet, paradoxically, the delphinium is also a cottage-garden flower, grouped in the folk memory with hollyhocks and honeysuckle and foxgloves, wholesome, even a little sentimental, a picture-postcard plant to complete the front gardens of thatched cottages. I have a book with just such a front garden on its jacket; in the foreground bright blue delphiniums, not especially carefully staked and even beginning to set seed at the bottom, lean slightly sideways among madonna lilies and look irresistible.

So in our garden, in front of our thatched cottage, there is no need to deliberate on the delphiniums' neighbours. Green leaves will do, anything will do so long as there is no empty space. Along our cottage path this hot July there were *Alchemillla mollis* and silver artemisia, pink valerian and purple Canterbury bells. On the other side of the border, between the delphiniums and the grass, there was a chance accumulation of things: a few dull, rose-coloured sidalceas, a satisfying spread of *Salvia* × *superba*, violet trimmed with red; the rising flat heads of *Achillea filipendulina* 'Gold Plate' were still young and green rather than gold; the domes of *Echinops ritro* were still silver rather than blue; there were one or two metallic, starry globes of *Allium aflatunense*, and at the back, higher than the delphiniums, climbed the deep pink roses. Everything had its own reason for being there, though nothing was there because of the delphiniums. They were thickly massed and all in flower – it was enough.

LARKSPUR & LOVE-IN-A-MIST

Consolida and *Nigella*

Now, in July, with the garden burgeoning, come the annuals, larkspur and love-in-a-mist among them.

Larkspur must be sown where it is to flower; it dislikes a move. Mine was sown last year and most of it was devoured by rabbits. A few seedlings overwintered, to grow 60cm/2ft high, for they were the tall, branching, single variety. There was a washy-white and a dusty-pink, but most of them were intense blue – the remembered blue of the medieval stained glass in Chartres cathedral and, like the glass, cut into small shapes, lozenges and diamonds, buds and pointed sepals. A more solid ration of that blue would be too dazzling for the eye. But the whole plant looked light and starry; even its leaves were fine-cut and insubstantial.

The flowers are recognizable delphinium flowers, but much smaller and quite without the delphinium's opulence. Close up they are like diagrams devised to explain, to beginners, the complexities of the delphinium flower. Here you can see clearly how the eye-petals, the nectaries, fit into the fly-away spur which curves up from the topmost sepal in a lyrical arc at the back. The sepals are five in number, the pale eye-petals appear to be only two, folded as if an origami expert were demonstrating a swan: there are two necks curving up to form points, and two tails disappearing into the spur.

But the flower commemorates the lark, not the swan. 'Larkspur' is only one of its common names; according to Parkinson, it was once called 'Larkes heeles' or 'Larkes toes'; but always it was that long spur at the back that inspired the name. It has been grown in English gardens since the Elizabethan age, and the tall single form spread into the cornfields, like the pheasant's-eye adonis, and became a weed. Again like the adonis flower, it spoke of a mythological death, and was supposed to have sprung from a young man's blood. Observers fancied they could read, in marks on its petals, the sigh of the earth ('Ai!') when Ajax turned the sword on himself after attacking a harmless flock of sheep. The species became *Delphinium ajacis*, the flower of Ajax, and was the first of the genus to reach England from abroad.

Yet now it has been given a different generic name: *Consolida. Delphinium ajacis* has been swept away: it is now called *Consolida ambigua*, a name which might well reflect the blushes on the botanists' cheeks. Likewise the native species which was once called *Delphinium consolida* is now called *Consolida tuntasiana*. This puts an unwelcome strain on the memories of gardeners. *Consolida*, however, means to heal or to fill. The larkspur once had a reputation for all manner of healing and soothing properties; the juice of larkspurs was an aid to improve failing sight, to ward off body vermin and to repel scorpions and snakes, so the name does not fit too badly, and in any case everyone will continue to use the common name, larkspur, as before.

If you are choosing larkspurs from a seed catalogue, you are faced with a different set of names. There will be the 'hyacinth-flowered' (both

ABOVE: This imperial larkspur looks like a cross between a delphinium and a double clematis with broad, pointed sepals.

RIGHT: In *Nigella* 'Miss Jekyll' a pattern of points is built up with overlapping sepals, central carpels and involucre bracts cradling the flower.

'dwarf' and 'giant') and the 'stock-flowered'; there will be 'tall rockets' and 'dwarf rockets' and 'Giant Imperials'. It does not matter to you whether the hyacinth-flowered and the rockets belong in the *C. ambigua* species, and the 'Imperials' in the *C. tuntasiana* species (some are almost certainly hybrids between the two); what matters is whether you want tall ones or short ones, doubles or singles, straight or branching. The hyacinth-flowered larkspurs are blunt at the tops of their spikes, as the name suggests; compact, non-branching, and good for gaps at the front of the border in their 'dwarf' forms; the stock-flowered type, unsurprisingly, are like double stocks, though tall; and the taller 'Imperials' may be the descendents of Parkinson's favourites, which he described as 'the double kindes . . . which stand like little double Roses . . .'

My preference is for tall, single and branching – perhaps because that is what I have got this year. They are listed in Suttons' seed catalogue as 'tall Rocket mixed, Height approx. 75 cm'. They have the wiry grace, the uncluttered look, of belladonna delphiniums; the central spike is not more important than the laterals. It is delightful to grow a tall plant which looks so fragile and emphemeral, which does not need staking and will flower its head off for weeks.

But that is the special contribution of annuals to the flower garden. Because they do not need to spend any energy feeding their roots and their only chance of immortality lies in setting seed, they tend to flower for longer than perennials. 'The garden Larkespurre', as Lyte wrote in 1578, 'floureth all Somer long'. I once sowed stock-flowered larkspurs in the spring in an awkward gap; by the time I had given them up, or forgotten about them, they suddenly started to flower, and with their soft pretty colours they continued for weeks and saved the border's life.

Gerard described larkspurs as 'sometimes of a purple, sometimes white, murrey, carnation . . .' 'Murrey' meant mulberry; 'carnation' perhaps meant clove-pink. A modern seed catalogue lists larkspurs in 'denim-blue'. There was a time, in the twenties, when the word 'larkspur' itself became a fashion colour. 'Larkspur', according to the *Daily Express* in 1927, was a 'pastel blue slightly inclining

to mauve'. All these colours, from the pale to the dark, from the pinks to the blues, can be seen in the tall bunches of 'Giant Imperial' larkspurs for sale during the summer on flower stalls at street corners. They have a steady sale, for they are famously co-operative cut flowers – better than delphiniums, whose hollow stems need cutting under water if they are to last. Larkspurs make favourite dried flowers, too, if they are picked before maturity and hung upside down in a cool, dark place.

All the seed catalogues carry a warning: larkspurs are poisonous, both in leaf and seed. Otherwise, they present few problems. The seed can be scattered, or sown in drills, directly into the garden soil, either in September or March. Enriched soil suits them better than poor, and autumn sowings supply better flowers than spring. They like full sun and, uniquely in their family, can put up with drought. But the more you water them, the taller they will grow.

At first, you would not believe that love-in-a-mist and larkspur are relations, unless you looked at their leaves. If the larkspur's leaves are fine-cut and insubstantial, the leaves of love-in-a-mist are as feathery as fennel – indeed, one species, *Nigella hispanica*, used to be called 'the fennel flower'. Perhaps annuals do not need substantial leaves because they are not storing up nourishment for next year; the leaves of the annual *Adonis aestivalis* bear this out, finely divided and curly as an endive, with an underlying crowsfoot structure. All three annuals of the Ranunculaceae family have the typical pedate leaf-division, but their leaves lack the flesh on the leaves of their perennial relations and seem like thin green skeletons. Of the three, the most important are the leaves of the love-in-a-mist; they are an essential part of its character.

They start just underneath the flower itself: as Gerard said:

> every floure hath five small green leaves under him, as it were to support and beare him up . . .

In other words, it has an involucre, recalling the winter aconite of January and the anemone of March. Five feathery bracts curve up protectively round the flower, soft as down when it is in bud, stiff as bristles when the seed-pod is forming.

The flower lurking inside the involucre's cup is round but also pointy. It has five coloured sepals (which comes as no surprise by now to the student of the Ranunculaceae) and each one is sharply pointed at the tip. In semi-double varieties there is a second ring of sepals lying flat upon the first, and overlapping them as in a semi-double delphinium floret. A pattern of points is built up, echoing the pointed tips of the involucre bracts and the five high, pointed horns of the central carpels; these rise prominently above the stamens, as green and threadlike as the involucre, and reach out, radially, like the spokes of a wheel. Indeed, the whole flower has this radial movement which explains why it was sometimes called St Catherine's flower, in memory of the martyr's wheel. Between the sepals and the carpels there are five very small petals or honey-leaves, bent and cleft at the tops with a hollow nectariferous claw below. But unless you catch the flower young, you will not see them; they soon fall, and the sepals linger on. When the sepals fall too, it is the turn of the seed-pod to take over.

This seed-pod is another essential part of the plant's character. It puffs out into a great plum-coloured airship surrounded by ferny whiskers. It crackles if you pinch it and is finished off with four to six outward-reaching bracts like green antennae. Inside are the black seeds which give the plant its Latin name: *nigella* is the diminutive of *niger* – black. Parkinson was annoyed that these seeds could not be used medicinally or in cooking, unlike the aromatic seeds of *Nigella sativa*, the nutmeg flower, which has disappeared from modern gardens. Altogether there are sixteen species of nigella, but only three are offered in the seed catalogues of today. There is the Spanish species, *N. hispanica*, which has large, single, mid-blue flowers, 60cm/2ft tall, deep pink stamens and a maroon-black centre, recalling certain anemones and ranunculus; there is a yellow-flowered oddity called *N. orientale* 'Transformer', whose seed-pod, according to Mr Fothergill's catalogue, 'can be turned inside out to form what appears to be a silvery-green and rather exotic, tropical flower'. And then there is *N. damascena*, our familiar love-in-a-mist itself, thought to have come to England from Damascus and now offered in several varieties: 'Persian Jewels' (mixed white, pink, mauve and purple), 'Miss Jekyll' (semi-double, sky-blue and 50cm/20in tall), 'Miss Jekyll Alba' (a pure white strain) and 'Dwarf Moody Blue' (exclusive to Thompson & Morgan, an edging plant 15cm/6in tall). This last would have put Miss Jekyll herself into a bad mood: 'One of the annuals that I think is entirely spoilt by dwarfing is love-in-a-mist,' she wrote, 'a plant I hold in high admiration.'

Because it is so light and lacy, one tends to think of love-in-a-mist as short. In fact, the 'Miss Jekyll' variety is tall, not to say lanky, and flops about unless it is propped up. It looks best among other things on which it can lean, inconspicuous and seeming to occupy no space at all until it opens its sky-blue flowers,which will fit in wherever it finds itself. Sometimes I grow it among 'Doris' pinks, a combination of babyish prettiness. This year I sowed it in big, oxidized coppers where *Lilium regale* grows; it filled in round the edges with ferny green, as well as leaning out recklessly over the edges, and when the lilies were over, it mitigated their loss with its innocent blue flowers. But it is a prodigal self-seeder and is likely to appear in unexpected places: in the middle of the creamy 'Dunwich Rose', or mingling with *Alstroemeria* Ligtu hybrids. The places it chooses for itself always tend to be particularly good.

It invites epithets like 'babyish' and 'innocent'. I have seen it recommended as a seed suitable for children to sow in their gardens, among those standbys, nasturtiums and cornflowers and candytuft. It is not fussy about soil; it likes sun but will tolerate a little shade, and should be sown where it will flower. Like the larkspur, it flowers best from late September sowings, but if it is sown in March, it will be flowering by July. It is a good cut flower, a welcome blue dried flower in pot-pourri and an amusing flower-shape to press. As a child I much preferred it to the acrid-smelling nasturtium, though I expect in the end it was its name that caught my fancy.

There is something about the way its flower is glimpsed through a netting of leaves that has invited a string of descriptive pseudonyms: 'bird-drinking-at-a-fountain' – a reference, perhaps, to the circle of erect carpels in the middle of a pond of blue; 'prick-my-nose', 'lady-in-the-bower' and, in France, '*barbe-bleu*'. To some imaginations, its secretive flower among spiky defences has suggested wickedness and it has become 'Devil-in-a-frizzle' or 'Devil-in-a-bush'. More often, it has represented a sort of Sleeping Beauty beyond the briars, and has been given amorous names: '*cheveux de Vénus*', 'love-in-a-puzzle', 'love-in-a-hedge'. But in English, the name that has stuck has been the softer 'love-in-a-mist'. The mist is the ferny ruff; the love is the exquisite, luminous blue.

MONKSHOOD

Aconitum

In August, despite continuous sunshine and yellow patches in the grass, a general darkness settled on the garden. It was the darkness of full-leaved trees whose fresh greens had gone and of my spirits as I numbered off all the brave plants that had finished flowering. And the August representative of the Ranunculaceae was dark, too.

Dark, dark blue with dark green leaves, my monkshood (*Aconitum napellus* 'Bressingham Spire') seemed to absorb light and did not reflect it. Unlike the delphinium, it sent no messages across the lawn; it did not sparkle or vibrate or seek to catch the eye. The peacock butterflies, dancing in the air, zigzagged past it to settle on delicious mauves of buddleja and catmint. John Raven thought the common monskhood, *Aconitum napellus*, 'dingy and lifeless'. Graham Stuart Thomas thought monkshood looked better planted in shade than in sun; shade, he observed, intensified the 'evil beauty' of its colouring. I had always thought of monkshood as belonging in the category of dear old things. 'Good old monkshood,' I would say. Yet it attracted these unkind epithets from garden writers.

It is not just a question of its colour, but of its shape. Hoods are menacing, and the distinguishing feature of all monkshoods is the cowl, or hood, which the topmost sepal makes to protect and conceal the nectaries and stamens within. Parkinson describes the flower's structure graphically as:

> composed of five leaves, the uppermost of which and the greatest, is hollow, like unto an Helmet or Headpeece, two other small leaves are at the sides of the Helmet closing it like cheekes, and come somewhat under, and two other which are the smallest hang down like labels, or as if a close Helmet were opened. . . .

Some people have seen it more benignly as Venus's chariot pulled along by doves, but to see this image you have to detatch the top sepal and hold the flower on its side. Whatever way you hold it, it clearly takes its place within its family. Here are five coloured sepals, hidden inside which are numerous stamens and five folded petals, so small as to look almost like stamens themselves. The two uppermost petals are clawed and prolonged into the helmet or hood. The similarity of this flower structure to that of the aquilegia and delphinium and larkspur is clear, but the difference is clear too: the joyous tapering spur of delphinium and columbine has been absorbed into the strange, secretive hood. Into this hood the bumble-bees work their way, and if you, too, approach it closely and tip it upwards to peer inside the flowers, you glimpse its secret black centre.

But my 'Bressingham Spire' does not invite close-up viewing. It is not an individual, but a crowd. It spreads and spreads so that even the

Aconitum napellus 'Bicolor' has large white flowers smudged with violet-blue on the lower sepals. The buds on the branching laterals are chartreuse-green.

separate stems are lost in a solid block of blue. The great point about 'Bressingham Spire' is that it has a cluster of secondary spikes tightly packed against each central stem – side turrets below towers, which is why its season continues for so long and why its effect in the border is solid blue. Its breeder, Alan Bloom, proudly describes it as 'perfect in form and strong enough to resist a tornado'. Yet there is a snag; the central racemes turn to seed all too quickly, often seeming to die off from the tops downwards, turning yellow, then brown. Sometimes a whole stem turns colour too. (I suspect this happens most in dry seasons, for aconitums, like the rest of their family, like moisture.) But the secateurs must be at the ready if the clump of monkshood is to remain handsome.

It is easy to draw up a balance sheet listing the pros and cons of the plant. On the debit side I am tempted to put, first, its appearances of dying at the top; though the lower lateral blooms may still be in perfect shape, they cannot distract the eye from the fact that the tall central raceme is seeding, making the whole clump look half-over when it could equally be considered as having only half-begun. Second I put its spreading habit (though this point may reappear on the credit side later on). But as it spreads, it weakens itself. Over the years, the stems grow shorter, the flower-spikes poorer, and the good gardener will dig the whole colony up every third or fourth year in the autumn or winter and throw most of it away. The best bits will be replanted in enriched, moist soil; but if the gardener should chew a leaf while he is doing this, or mistake a stem for a bit of celery (as the Victorian lady gardener, Mrs Loudon, thought 'ignorant persons' might do), he will meet the fate of those described by Gerard whose:

> lipps and tongue swell forthwith, their eyes hang out, their thighes are stiff, and their wits are taken from them . . .

In short, he will be poisoned.

This is the third point on the debit side of my monkshood balance sheet. Every writer on monkshood mentions the deadly poison that runs through the whole plant, particularly its roots. Its Greek name was given to it by Theophrastus, perhaps as a warning that it was used to poison darts (*akon* – a dart). All aconitums contain a poisonous narcotic alkaloid; a fable has it that they sprang from the saliva of Cerberus. But the chance of eating the aconitum's roots is so remote that I cannot count this as a serious drawback; it is not a bulb that could be mistaken for an onion; it is a small, unappetizing tuber.

So now for the credit side of the sheet. It is a tough plant, described as 'bone hardy', and as 'an easy plant for the lazy gardener'. It will grow in sun or shade; it spreads, which makes it both economical and generous (as well as a nuisance); though it is quite tall, it does not need staking; and best of all, it flowers in August. It keeps at bay the gardener's August blues, by being blue itself.

All aconitums have dark, glossy foliage, easily recognized and pigeonholed by the Ranunculaceae fancier. It is five-partite, the segments lobed and toothed like the foliage of the trollius or delphinium, though more deeply divided than they – the slits go right down to the base of the leaf, and the leaves clothe the whole flower-stem from the tip to the base. Like the delphinium, it comes early through the soil in spring, and makes rounded, prosperous-looking clumps which grow with satisfying speed.

There are three hundred species of aconitum, but only one or two or them get into gardens, and these are mostly varieties of *A. napellus* or hybrids between *A. napellus* and another species. The classification of aconitums is so perplexing and inconsistent that it seems safer to give only the varietal name and omit the botanical name altogether. I have five other varieties in my garden besides the famous 'Bressingham Spire'. The first is called 'Newry Blue'. I have only dug it up and divided it once; it fell easily into crowns, some few of which I replanted; the rest I tipped on to a rubbish heap. The indestructible thing pushed through the heap and flowered there the following summer. It is a quintessential monkshood, hardy, vigorous, spreading, deep blue, with branching leafy stems about a metre high, but its distinguishing feature is that it starts to flower in June, and keeps going for two months.

Next comes a hybrid called 'Ivorine'. This is quite distinct from the *A. napellus* varieties, and is usually called *A. septentrionale*. It does not spread, its roots are not tuberous but fibrous; it is not very good at holding itself up straight, and it is best admired in close-up. As its name implies, its flowers are not blue, but ivory – the colour of new ivory, tipped with green. They are small, less like helmets than exaggerated little cloche hats from the 'twenties. There is a central raceme and wide-branching laterals whose flowering is paced out over many weeks, sometimes first opening in May, sometimes still flowering at the end of August. Such charming generosity should make it a star of the flower border, but there is something about it so pale and self-effacing that often one does not notice it is there at all. This year mine hid itself among the powder-blue umbels of the sea holly, *Eryngium × tripartitum*, which embraced it usefully in stiff, prickly arms but entirely eclipsed it with superior beauty so that even the bees ignored it, clustering obsessively all over the sea holly's mist of blue. You would think that ivory and powder-blue would enhance each other; in fact, this gentle ivory would probably be enhanced by an even gentler background of variegated leaves.

A much showier hybrid of *Aconitum napellus* appeared in August: *A.* 'Bicolor', whose large flowers are white smudged with violet-blue. It is as if the blue were not a fast colour, and had run down into the lower sepals, leaving the high hood milky-white. While the central spike flowers, the buds which wait their turn on the wide branching laterals are fresh chartreuse-green. The result is a happy monkshood in party mood. I enjoy it; it actively enhances and lightens my border, yet I regard it as a sport, for I cannot relinquish the belief that the monkshood's rightful mode is funereal.

For this reason I rate my fourth monkshood, 'Spark's Variety', most highly. The blue is deepest midnight, a light navy-blue with violet in it. The flowers are large, with perfect aquiline profiles. They have a sombre, queen-of-the-night beauty. But the best thing of all about it is the way the flat, dusky buds are held out on the laterals, quite a distance from the main stem, the whole flower-head making a large triangle.

Finally comes a late, September-flowering monkshood, *A. carmichaelii* (formerly *A. fischeri*). In some forms it is said to be a comparatively pale blue, but mine is deep violet.

All the dark monkshoods look their best with the yellows and oranges of late summer. I have seen 'Spark's Variety' growing through a brilliant crocosmia – indigo-violet hoods branching among sprays of fiery orange. You could achieve the same effect by planting monkshoods among arching orange lilies, *L. henryi* or *L. tigrinum*. Graham Stuart Thomas recommends planting *A. carmichaelii* with the September-flowering red-hot poker, *Kniphofia triangularis*. Other writers have other tastes: one likes his monkshoods mixed with another sombre, hooded flower, the acanthus. In my garden they stand in front of shining pink Japanese anemones. I conclude that they are like delphiniums; they will never clash with other plantings and belong anywhere. They supply the shadow that accompanies the light.

AUTUMN

JAPANESE ANEMONES

Anemone japonica

The lightest September flower is the Japanese anemone. It looks positively weightless poised high above the herbaceous border on the tops of straight stems that never bend or tilt. If it is silver-pink its face shines in the light; if it is white it illuminates dark corners. It will grow and spread and flower in full shade, planted between shrubs. But it flourishes in full sun, too.

The white is always pure white, the pink is always cool pink with a trace of mauve in it, sometimes paler, sometimes deeper. There are no other colours, but there are many variants of shape, size and height, in a steady stream of new named cultivars. Some are relatively short; some (like 'Luise Uhink') are semi-double with thin, slightly fussy sepals giving the flower the look of an aster, and in some, the sepals are pinched at the base like the petals of an osteospermum; some, like 'White Giant', have big flowers 10cm/4in across. But in the world of the Japanese anemone, a single, saucer-shaped flower on a tall stem looks best.

Pink Japanese anemones in September.

Among the whites, the old hybrid 'Honorine Jorbet' (sometimes called *A.* × *hybrida* 'Alba' or *A. japonica* 'Alba') is peerless; it is 100cm/40in tall, and its simple flower has five or more white sepals and a ring of golden stamens. It has adorned shady garden borders for a century, rising above hydrangeas and dull evergreens, robust in growth but delicate in detail, flowering for three successive months, almost too familiar to be showy, too obliging to be adequately cherished.

Among the pinks, 'Queen Charlotte' was E.A. Bowles's special favourite. 'It is so lovely', he wrote, 'that I cannot bring myself to root any of it out.' It is not tall and has large, generous flowers, tip-tilted semi-double saucers. But my pink is called 'September Charm'; it is the pink counterpart of 'Honorine Jobert', single, tall, long-lasting.

It does not wait until September to open; I first observed it closely in August this summer – which was just as well, since by September it was not itself. Thirst had made it dwindle; its leaves had become dull, its flowers, still barely held aloft, had shrunk in size. It is the easiest of plants to grow, but it does like moisture.

When September arrived, the garden was bone-dry. Throughout July and August there had been heat and little rain. There were casualties everywhere, drooping leaves, browning stalks. Day followed windless day of blue skies, warm sunshine; the apples fell early and small from the trees and the blackbirds scuttled about pecking at them; dried leaves lay beneath the crab apple and crackled depressingly underfoot; the air was a-dazzle with butterflies. Non-gardeners sat about in deck chairs or snoozed on the straw-coloured grass. This was the very moment, in early September, when Keats wrote his 'Ode to Autumn', celebrating the suspended feeling that 'warm days will never cease'. I grimly carried cans of water to my newly planted specimens of 'Luise Uhink' and 'Honorine Jobert'. Their leaves had shrunk and darkened so much that they barely showed; there was no hope of observing their flowers at first hand this year.

But there was scant hope anyway. Japanese anemones always start by sulking, and seldom think of flowering until they have lived for at least two years in a new place. But once they perk up, there is no stopping them. They grow in all directions, horizontally and vertically. Before long the small original plant is filling a square metre of border space and a dozen or more straight stems are rising from the handsome basal leaves until in August they are waist high.

The plant grows from a tuberous rootstock which spreads out horizontally 20cm/8in below the ground. In spring it makes strong clumps of divided leaves, dark green on top, pale green beneath, incised round the edges and arranged in the typical pattern of threes. The strong brown stems branch and branch again, four separate flower-stems coming from a single stalk, and each of these flower-stems branches again into four or six, and finally there will be four to six flowers, each on its separate pedicel at the end of each stem. The seed-heads are so round they look like buds, but they are yellow, and the buds are mushroom-pink. The final effect is a mixture of buds, seed-heads and open flowers all enhancing each other; the sequence goes on and on, from bud to flower to seed-head, in a cycle that continues from August to October.

The open flowers are composed of coloured sepals encircling a central boss of lavish stamens – we are back in the simple world of a child's drawing of a flower, like the marsh marigold or the winter aconite. The colour of 'September Charm' is not constant; nor is the number of its sepals, which varies between five and ten. At the back of the flower there are usually three sepals stained a deeper purple-pink – it is the same lilac-rose colour as the buds, and if the flower stands against the sun this deeper colour shines through so that the silvery, pearly pink of the front shimmers in contrast with something deeper. And then there is the contrast of the frosted pink sepals with the warm yellow ochre of the stamens, and of the yellow stamens with the pale green button in the centre of the flower. It is all contrast, but no clashing.

Because pink and yellow are already mixed within the flower, it is easy to mix 'September Charm' with the bright yellow border flowers of late summer, the heliopsis and rudbeckia and helenium. Its cool company is exactly what these bold daisies need.

But Japanese anemones, like delphiniums and monkshood, larkspurs and love-in-a-mist before them, look good anywhere. In my garden, mixed with *Aster × frikartii* in a sunny border, they bring a pastel pink and blue softness to late summer. It would be good to try them in the shade beside another tall, lilac-pink September flowerer – *Eupatorium purpureum*, the hemp agrimony – just as it would be good to plant white Japanese anemones beside the white shrub rose 'Nevada' to catch its second blooming. It is hard to imagine a garden that had no place for them, or to imagine a gardener who could not grow them. Enthusiastic gardeners can mulch them and discipline their exuberance; Miss Jekyll divided hers every October using a plasterer's hammer with a cutting edge. But unenthusiastic gardeners can safely neglect them; they look after themselves; they never need staking and don't like being moved about. All they need, once they are planted, is a clear-up of the old stems when they die; this can be done in the autumn with secateurs, or in the spring with the hands, for by then the stems are dry and brittle and break off at a touch. In March new growth is pushing strongly through the earth, much at the time when the soft leaves of *Anemone blanda* are beginning to show.

I used to think the name 'Japanese anemone' was some sort of quirky metaphor. It never occurred to me that the trusty, tough plant standing tall in the late-summer herbaceous border was the same genus as the tiny, frail anemone of spring. But now I accept that it is: it grows from a tuberous root system; its leaves may be tough, but they are cut in much the same pedate shape as the soft, wine-brushed leaves of *A. nemorosa*; and the flower, of course, follows the same pattern: five or six coloured sepals, some of them smudged with rose-colour on the backs, no petals, stamens in a lavish ring round a neat, domed centre. To clinch matters, the scented woodland anemone, *A. sylvestris*, which flowers from April to June, is so like a small and refined replica of 'Honorine Jobert' that it can be mistaken for it.

Like *A. blanda*, *A. japonica* is a capricious cut-flower, sometimes happy, sometimes not. It is more likely to last in water when its stem is cut short. It may not come from Japan (its approved botanical name is *A. hupehensis*, which means that it comes from the province of Hupeh in central China), but it really *is* an anemone, which explains the simple beauty of its face.

BUGBANE

Cimicifuga

September is half over and there is a chill in the early morning air and slanting light in the afternoons when *Cimicifuga racemosa* flowers. There it is at the back, towering like a small tree over the deadheaded border, rescuing it from decay. Two metres tall or more, *C. racemosa* rises and then bends its flowers very slightly forward over its dying companions. As its name implies, the flowers come in racemes, sometimes 30cm/12in long, white, fluffy and exceptionally thin.

It is the thinness of the cimicifuga flower-spikes which sets them apart. But the structure of those inflorescences recalls the thalictrum's flower. It has no petals worth speaking about, but its five sepals are an essential part of its beauty, for they hold the buds into tiny, perfect globes like beads, spaced out and distinct along the flower stem. The separateness of each bead-bud from its neighbour is as important to the total elegance of the plant as is the spacing of the buds at the top of a delphinium spike. In some varieties the buds are tiny balls of lime-green; in one (*C. ramosa* 'Atropurpurea') they are purple, and in one (*C. simplex* 'Elstead') they are pink. When the buds open the sepals, like those of the thalictrum, are destined to fall, giving way to a tight-packed cluster of white or cream stamens which fan out to touch their neighbours so that what was once a series of distinct buds becomes a continuous inflorescence, a bottlebrush of anthers. The process is gradual; for some weeks the discreet buds are all that can be seen, tapering up to the tip of the raceme; one day, the lower buds begin to open but at the top of the spike they are still tight shut. When finally the opening movement reaches the top, the lower bunches of stamens have lost their colour and have become stiff little seed-heads held at right angles to the stem. The whole process extends over several weeks, and even when the leader, the central flower-spike, has completed its cycle, there are as many as ten lateral spikes taking their turn to flower below. It is the mixture of shapes and shades, the combination of fluffy anthers and tight round buds, followed by neat seed-heads, that gives the whole plant its allure. Its movement is upwards, not outwards at all; even the lateral spikes are part of the upward thrust.

Tallness, as well as thinness, sets the cimicifuga apart. Its stems are like canes, so strong, despite their slenderness, that they carry themselves erect and self-reliant even at well over head height. The height seems to come from the length of the internodes – just as, in a very tall boy, it seems to come from the length of leg between hip and knee, or between knee and ankle. The cimicifuga's nodes, from which the leaves grow, are spaced as much as 60cm/2ft apart up the stem, and after a midway spread of proudly held compound leaves, the leafless stem soars on up to the bare inflorescences, held high above.

I saw a beautiful purple cimicifuga in someone else's cramful cottage garden – it looked like a long-

Cimicifuga ramosa: a towering autumn plant, whose honey-scented bottle-brushes arch and loop on their strong, thin stems and attract late butterflies.

legged bird. There were 60cm/2ft of stem before the serrated and segmented leaves spread out laterally like the fat feathered body of emu or ostrich or peacock; above them, the stem continued like a slender neck. The flowers were not yet open; the buds were still tight beads of dusky mulberry colour, exactly the same colour as the stem. It was as if the whole plant had been dipped in dull purple dye, but the colour did not take completely on the leaves, where a little olive-green showed through.

Then I saw the same plant, *C. ramosa* 'Atropurpurea', as a beautiful container-grown specimen in a garden centre, purple all over this time with no hint of green, and with a giddy price tag tied to it. Obviously it is a collector's piece, but I am busy planting it in my imagination, now here, now there in the garden.

It is a challenging plant to place. Its dull, dark richness (for *atro-purpurea* means black-purple) is not enhanced by ordinary surrounding greens. It would look better against sage-green, or silver-green, or grey. There is a temptation to adopt the fashionable idea of growing it near other reds and purples, cut-leafed Japanese maples perhaps, or with small purple plants at its feet, *Heuchera* 'Palace Purple' or *Sedum* 'Ruby Glow', but best of all I like to imagine it planted by itself on a garden promontory where its whole shape, from soil level to tip, would be surrounded by simple space – air – on either side.

Meanwhile I have my own long-loved, long-lasting cimicifuga which is no trouble at all to place and will mix with anything. It is *C. rubifolia*. For years I called it *C. racemosa cordifolia*, and for years it has soared up near the front of my herbaceous border, a north American species with broad-lobed leaves like a glossier, more handsome version of the Japanese anemone's. These leaves form substantial basal clumps as well as branching out halfway up the stem. The stems are dark and slender, but strong as tree-trunks. The sepals, holding the tiny buds tight, are the colour of coffee cream. The central spikes are at least 30cm/12in long, and when they start to die the colour-scheme is simply reversed, cocoa at the bottom and cream at the top. The mixture is delectable, but it is not showy. An aristocrat with perfect manners, it does not need to push itself forward or attract attention; it is politely complementary to its neighbours, though it never really mingles with them, always holding itself slightly aloof, and it may be some years before you come to the conclusion that it is more important than they are. It will always be a contrast to its neighbours which is why it complements them so well.

It flowers from early August to mid-September, while many more common herbaceous flowers around it come and go. At first the great smoke-blue globes of the echinops are beyond it for the brief moment of their flowering in my garden, and in front of them the cimicifuga flower-spikes look more delicately frail than ever. It has *Lavatera* 'Barnsley' as a companion for weeks and weeks; they could not be more different from each other, except in length of flowering: the lavatera, plump and pretty, pink and white, a generous, blowzy shrub bringing the background to the foreground, and the fastidiously self-contained cimicifuga standing in front of it but seeming to shrink to the background – a star taking a supporting role. Round its skirt of dark leaves spread a silver artemisia, 'Powis Castle', the pink-flowered cranes-bill *Geranium sanguineum lancastrense* (*G.s. striatum*), and the flat, fleshy heads of *Sedum spectabile*; it indulges them all. A few metres farther down the border another tall plant with white racemes is growing: *Veronicastrum virginicum*, which at first may seem uncomfortably to compete with the cimicifuga's stance. But it cannot really compete. The veronica is straight white mixed with straight green – no subtle gradations here; it is softly pretty but lacks the cimicifuga's mysterious, long-lasting substance and soon it is flopping forward, calling for help.

I never thought of picking my cimicifuga and taming it for the house until, in a drought-stricken August, I needed tall flowers for a family wedding. Then, in the early morning, I cut the towering stems; the secateurs sounded as if they were snipping through wood, I split them and plunged them into water. Later, they lay in the car, reaching right through between the front seats from tailgate to dashboard. They survived the perilous journey and adorned their arrangements with thrilling rarity. Afterwards, if they died, they did it so quietly that no one could tell; they just seemed to stiffen a little more, and their racemes, still tapered and beautiful, were dry to the touch.

The common name of cimicifuga is bugbane. *Cimex* means 'bug'; *fugio* means 'to run away'. The plant has been known in England since the seventeenth century, when its leaves were used to ward off fleas. As late as the Victorian age, William Robinson complained of its 'odour'. In his *English Flower Garden* he dismissed it as 'not of much garden value', though he allowed his readers to contemplate it for the wild garden, planted in groups.

But the nose for plants changes, like the taste. Today, the cimicifuga is found to have a honeyed scent; the red admiral butterflies pose for their photographs on its fragant flowers; the catalogues of adventurous nurseries now list two or three species (often *C. ramosa* 'Atropurpurea' – sometimes named as *C. simplex* Atropurpurea Group – and *C. simplex* 'White Pearl', this latter desirable because it is the last to flower, not opening before October); but you are unlikely to find it listed in the index of any gardening book published more than ten years ago. It remains, for the time being, a 'plantsmanship' plant. If you refer to your cimicifuga you may watch a slightly glazed expression cross your companion's face. This is because of the plant's reticence, its refusal to be showy and catch the eye, and not because it is difficult to grow.

It is easy to grow. Not only does it never need to be staked, it never needs to be divided either, though it can be increased, by division, in early spring. It just quietly establishes itself and settles down for a long life. It likes water, and humus – not impossible demands to fulfil by anyone with a compost heap and a watering can. It likes shade, but it doesn't mind sun. So long as the soil is reasonably friable, it will grow well whether it is acid or lime. It is impervious to frost.

Autumn is all very well, if you have sedums and fuchsias and michaelmas daisies, cherry trees to change colour, a soft elegiac light blurring the edges of things – and Japanese anemones. But still to be waiting for just one last, tall plant, *Cimicifuga simplex*, to open in October – that makes all the difference between resignation and hope.

BANEBERRIES

Actaea

I didn't even know how to pronounce it.
'Have you got an actaea in stock?'
'A *what*?'
'*Ac*-taea.'
The line was bad. She couldn't hear.
'How do you spell it?'
'A-c-t-a-e-a.'
'Sorry, I can't help you.'
'But it's on your list.'
'Could you spell it again?'
Then she said: 'Oh – Ac-*tee* – a! Yes, we've got *rubra* and *alba*.'
'Good! I'll have both. One of each.'

So we drove across a whole county and collected them.

Their flowering was virtually over. You could just see the thin remains of short racemes in creamish green. But the leaves were in excellent shape; there were spreading clumps of them on the moist slopes of the Paradise Centre in Bures, Suffolk, healthy in the rich leaf mould beneath the trees. I would have mistaken their ferny fronds for astilbes a week ago.

But I had just been reading R.D. Meikle's first book, *Garden Flowers*, where, unusually, the flowers are arranged systematically in groups

ABOVE: *Actaea alba*'s modest flowers will become extraordinary pink-stemmed cream berries later on. The divided, fine-cut leaves are typical of the Ranunculaceae and, spreading in moist shade, they make an attractive alternative to ferns.

RIGHT: The shining red berries of *Actaea rubra* set it apart. The actaea is the only member of the Ranunculaceae family to have berries.

according to their families. There they all were together, my friends the hellebores, anemones, aquilegias . . . but slipped in between the aconitum and the cimicifuga was this fresh one, *Actaea spicata*. It was like the twist in a family story where suddenly, just before the denouement, a stranger makes his entrance, and claims to be a long-lost brother.

Not that the actaea is cut out for making much of an entrance. It is the quietest character, a woodlander, a shade-dweller, whose racemes of fluffy white flowers are insignificant even when in full bloom, and short-lived. *Actaea spicata* is a wild flower and used to grow in limestone districts all over the temperate regions of Europe, flourishing in copses and moist places. William Robinson recommended it for 'rich bottoms in the wild garden'. It was called 'herb Christopher'.

But it has its own point to make. It bears berries, and it is the only member of its family to do so. That was what made me chase after it from Norfolk across Suffolk. The thought of rich berries – black on *A. spicata*, and red and white on *A. rubra* and *A. alba* – was irresistibly tempting; it would round off my story of the flowering year among the Ranunculaceae with a suitably autumnal offering.

Unfortunately, things did not work out as neatly as this, for my two actaeas set about producing berries in August, and by the autumn they had gone. Had the birds eaten them? And if so, had they died? These berries are deadly poisonous, which is why the plant's common name is the 'baneberry'. A chemist, in whose garden *Actaea rubra* had naturalized as ground-cover, told me it does not follow that a berry which would poison a man would also poison a bird. And the berries may not have been eaten, anyway. They may simply have withered in the hot, dry summer.

They started well. *Actaea rubra*'s berries changed quickly from cream to a delicious bright strawberry colour, each berry the size of a pea. *Actaea alba*'s berries remained peppermint-white, but their short pedicals turned raspberry-red and extraordinarily thick, a bizarre effect which made me wonder if there was something wrong with my plant. Each white berry was finished with a black dot; they evidently reminded someone of those white glass eyes with dark irises which used to rotate in dolls' heads long ago, and so poor *Actaea alba* got the pet-name of 'dolly's eyes'. Its alternative botanical name is *A. pachypoda* – much apter, for the Greek *pachsy* means 'thick' and here signifies the most distinctive feature of the berry – its swollen stalk.

The word *actaea* is said to derive 'from the Greek "aktaia", elder, the leaves resembling those of that tree' (*Plant Names Simplified*). The common elder seeds itself all through our wood, but its leaves are so much bigger and coarser than the baneberry's that the likeness does not strike me, though a fine-cut elder, *Sambucus nigra* 'Laciniata', comes nearer to the mark. But the leaves of *Cimicifuga simplex*, which grows nearby, are absurdly like their cousin's.

If I could grow only one baneberry, I would have to choose *A. alba* for its sheer audacity. Its bare stems rise well above its ferny leaves, and the white dolly's eyes with their black pupils are arranged in fat racemes with a raspberry core which cannot be ignored. Friends who pass it in our wood say 'What's *that*?' Then they smile and avert their eyes, but already I take pride in it. Next year, I shall top-dress both my actaeas with leaf-mould and keep them well watered until I am satisfied that their rhizomatous roots have established, and are about to colonize a stretch of wood. They looked at home from the minute I planted them side by side between a grass path and a spread of the perennial foxglove, *Digitalis* × *mertonensis*. Their height was just right at

60–90cm/2–3ft – neither tall nor short; their habit was right, the stems being slender but strong, the finely cut leaves spreading and soft; their colour was right, not dingy but fresh. They looked as if they had come of their own accord into the wood, Herb Christopher joining Herb Robert.

Next year, in the wood, it will be the actaeas for whose reappearance I am specially waiting. Meanwhile, there is the actaea in the botany book, and here its credentials are confirmed. Yes, it likes moist, well-drained woodland; yes, its leaves are ternately divided into threes, and sharp-toothed round the edges; yes, its flowers have five petaloid sepals, five to ten honey-leaves, and numerous stamens, though there is only one carpel to a flower and its seed is a berry.

But yes, it is without doubt a legitimate member of its family.

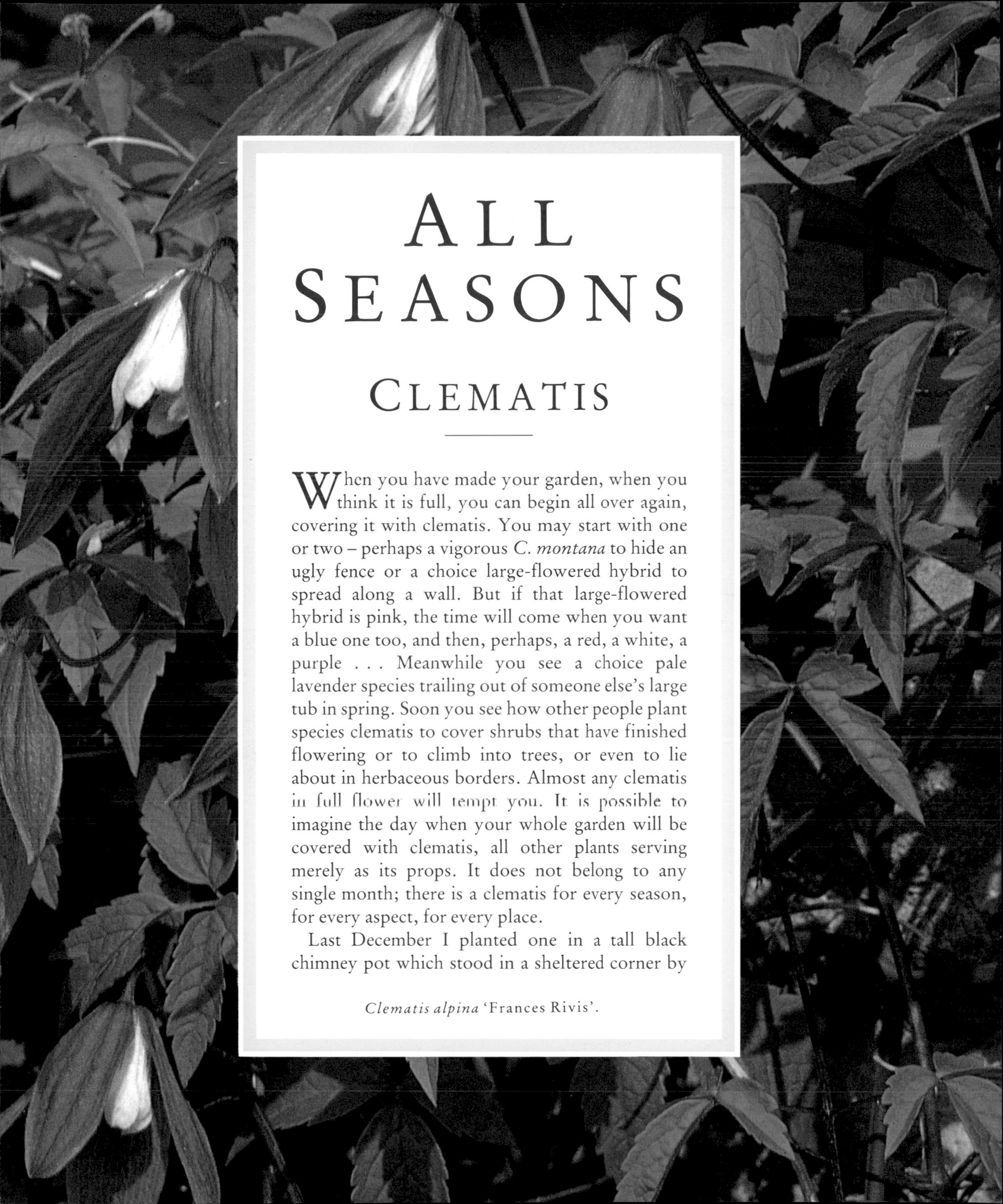

ALL SEASONS

CLEMATIS

When you have made your garden, when you think it is full, you can begin all over again, covering it with clematis. You may start with one or two – perhaps a vigorous *C. montana* to hide an ugly fence or a choice large-flowered hybrid to spread along a wall. But if that large-flowered hybrid is pink, the time will come when you want a blue one too, and then, perhaps, a red, a white, a purple . . . Meanwhile you see a choice pale lavender species trailing out of someone else's large tub in spring. Soon you see how other people plant species clematis to cover shrubs that have finished flowering or to climb into trees, or even to lie about in herbaceous borders. Almost any clematis in full flower will tempt you. It is possible to imagine the day when your whole garden will be covered with clematis, all other plants serving merely as its props. It does not belong to any single month; there is a clematis for every season, for every aspect, for every place.

Last December I planted one in a tall black chimney pot which stood in a sheltered corner by

Clematis alpina 'Frances Rivis'.

the porch. It was *Clematis cirrhosa* 'Freckles', fresh from the garden centre and in fully wintry flower. The flowers were pendent lampshades, as drooping as a fuchsia's; the backs of the sepals were a drained grey in tune with the December sky, but they were lightly tinged with heliotrope, and inside the sepals were spattered with wine-red spots, accounting for this variety's jolly name. It is not a jolly flower; it is as retiring and subtle as a spotted orchid and might not be remarkable in high summer but is treasured in December. Its home is the Mediterranean and it is not hardy, so I covered it with a curtain of plastic during the January snows and it emerged in spirited, spring-like mood, thrusting long tendrils out to left and right until it had wreathed a nearby window with its crisply divided, glossy leaves. This November, when an icy wind was blowing from the north, its leaves were as freshly bright a green as ever, but I went away in a hurry without covering it and when I returned those leaves were limp and sickly after frost. I am watching my patient anxiously; it has become the focus of expectation in my winter garden.

Clematis is called the 'Queen of Climbers'. It is also queen of the whole family of Ranunculaceae. Like all monarchs, it stands a little apart from the others and above them: it is the only member of the family that knows how to climb. You can buy it at any garden centre in the land. There will be a row of young, pot-grown plants tied to neat bamboos, sometimes upright, sometimes toppled over in the wind. One of them will probably be *C. montana rubens*; one will be 'Nelly Moser', and one will be 'Ville de Lyon'.

C. montana rubens flowers first – in May; it is enormously vigorous, 9 metres/30ft tall and just as wide; its flowers are four-sepalled, shell-pink, rounded as Tudor roses. It will perform as generously in shade as in sun, and does not need, or want, pruning, apart from trimming back when it outreaches its allotted space. The only word to be spoken against it is that it is *too* popular, too familiar to be a thrill. Why not, then, search out two very lovely and less common variations on its theme? One is *C. montana* 'Tetrarose', larger in flower and bronzer in leaf than the type; the other is *C. spooneri* (*C. montana sericea*), whose flowers are also larger but whose vigour is slightly less; in May it is a galaxy of waxy white.

Before the montanas have finished flowering, 'Nelly Moser' has begun. She is the familiar, rather vulgar lady with the bold carmine stripe down the centre of each pale mauve sepal; it is impossible not to notice her, which is perhaps why she is so often seen; or perhaps it is because she is reliably prodigal of flower, and will bloom with particular intensity in the shade. But there are choicer beauties in the same repeat-flowering, June/September group which are eye-catching, not for their stripes, but rather for the lack of them – for simple, concentrated colour on sensationally enormous flowers. They, too, are very famous, but just not quite as famous as Nelly. 'The President' has purple sepals with silver backing; 'William Kennett' is a hundred years old, and has pale lavender-mauve flowers, slightly crimped and crenulated where the sepals overlap; 'Marie Boisselot' is newer (in an earlier incarnation she was called 'Madame le Coultre'); she is pure white and tilts her flowers upwards towards the sun. All these early-summer varieties unfold their flowers first in May or June, and with luck, and light, judicious pruning, they will mount a second, less ambitious flowering in September.

'Ville de Lyon' belongs to the second half of summer – to the group that used to be called × *jackmanii*. They are the easiest to deal with, for hard pruning is easier than light pruning; you simply cut all of last season's growth ruthlessly away in February. 'Ville de Lyon' is red, and famous for the fact. But so is 'Mme Edouard André' – an ecclesiastical shade that is neither scarlet nor crimson but a sort of dusky puce touched by the colour of a bishop's amethyst ring. She is neater in growth than 'Ville de Lyon'; her flowers are 12cm/5in across, and the five sepals hold their shape as if they were starched; the edges curl up but the tips curve down, so each one is shaped like a lovely scoop.

It is a long time since I felt disturbed by the absence of true petals in the clematis flower. By now it is quite obvious that these shapely flowers have a particular body, a weatherproof substance, *because* they are made of sterner stuff than petals, stuff that can withstand weeks of wind and rain, the stuff that can hold a whole corolla tight-shut until it is time to open – in short, sepals.

Closed, the bud of a species clematis is globular, that of a large-flowered hybrid is long and pointed, often with a suggestion of concertina pleats. There is a silvery bloom on these pointed buds with just a hint of the colour which will later be revealed on the other side. Open, the points of the bud divide to supply the points of the star: sometimes six-pointed, sometimes eight, according to how many sepals there are; in the middle of every clematis flower is the burst of stamens that is characteristic of the buttercup family. The finishing touch to each different variety is the contrast between sepal and stamen: in my dusky 'Mme Edouard André', for example, the stamens are greeny-cream; in the deep purple 'President' the stamens are dark red; in pale mauve 'William Kennett' they are deep purple-brown, and in white 'Marie Boisselot' they are primrose-yellow. In every case the contrasting stamens give sparkle to the flower.

For there is no sparkle in the coloured sepals themselves; they have an opacity which makes it possible to construct clematis mock-ups out of card. In a big garden centre with at least twenty varieties for sale, I have seen these paper look-alikes attached to the bamboo stakes of the plants in November; they are sickeningly convincing, but at the same time they are enough to put you off buying any further large-flowered clematis for a season.

It is more encouraging to buy from a specialist nursery, where the sheer number of varieties (usually a hundred or more) is intoxicating. The pots stand in ranks in polythene tunnels, but a specialist grower will not sell them to you until they are two years old and have more than one stem coming from the base. If you do not live near a clematis nursery, there are hours of pleasure to be had thumbing through their helpful catalogues; your order can be posted to you in a long cardboard box from the middle of the summer onwards, for clematis resent root-disturbance and are always sold in pots.

Suddenly last year, after long inertia, I found myself ordering four new clematis – small-flowered species, to climb through other things. There was the buttercup-yellow *C. tangutica*, hung with countless Chinese lanterns and fluffy silver-grey seedheads all mixed together, which was to climb an ailing plum tree; there was *C. flammula*, the Virgin's Bower, to drape an old weatherboard privy with rampant stems of delicate, dark green leaves and myriad sprays of almond-scented flowers in August, each one a tiny white cross with a tassel of stamens in the centre; there was the herbaceous *C. heracleifolia davidiana* to plant between spent delphiniums, with china-blue hyacinth flowers and big coarse leaves like a Japanese anemone's; and there was *C.* 'Etoile Rose', whose ferny leaves and bell-shaped flowers were to weave through the thick stems of the Bourbon rose 'Mme Isaac Pereire', its sepals being much the same colour as the rose's petals, but each one fringed with silver and reflexed backwards like a Turk's cap. They all came with neat polythene bags over their pots, each bag secured to its bamboo stake with a twist. Clear planting and feeding instructions accompanied them.

My supplier was Fisk's Clematis Nursery. And I followed these new planting instructions just as, twenty-five years ago, I planted my first large-flowered hybrids with one finger in Jim Fisk's book *Success with Clematis*.

For a start, the planting hole must be 45 × 45 × 45cm/18 × 18 × 18in, which is quite deep, despairingly deep if you're digging into some never-cultivated spot, below a tree, which was probably once a path. The spade strikes flint and flint; the dry September ground is like concrete one foot down. You are tempted to cheat, and skip the last spit, or kid yourself that what looks like a foot-deep hole may really be rather more. When you've dug it you still have to loosen the subsoil and mix in some bonemeal. Don't forget, coaxes Mr Fisk, that you are making a home for a most beautiful plant which will live there for many, many years. You must then spread a layer of rich compost over the surface before infilling with topsoil or with newly bought John Innes No.3 if the old soil is

ABOVE: *Clematis florida* 'Sieboldii' is strikingly exotic, with greeny-cream sepals and a prominent boss of petaloid stamens.

RIGHT: *Clematis flammula*: galaxies of vanilla-scented flowers in lacy sprays, each one a tiny, four-sepalled cross.

poor. You must shake the plant out of its pot so that its roots remain intact (soak it in a bucket of water, perhaps, while digging the hole). But if any of its roots are spiralling round the earth ball, loosen them. If the plant has not made a firm rootball in the compost, then it is not a good one. Before you infill, you must remember to set an empty flower pot near it to come flush with the surface of the earth, as this will be a useful way of getting water to the roots later on. Weekly watering is part of the summer drill, with a spoonful of liquid fertilizer in the water if you want prolonged, prolific flowering, and you can regularly dribble this fertile water into the empty pot and it will filter through to the roots below. You set the new plant 10–15cm/4–6in below surface level, stripping the bottom pairs of leaves from the stem so that there will be submerged buds from which new stems can grow in case of the dreaded clematis wilt. Then you push the cane gently down through the potting soil to anchorage in the earth below, tipping the plant at a slight angle so that the cane points towards the tree or arch or wall up which the new clematis is to climb. Finally, you put a tile or stone over the roots to keep them cool, and resolve not to be disappointed if nothing much happens in the first year after planting; clematis take a time to settle down. Apart from the digging, planting a clematis is not difficult or skilled, just demanding. You are required to cosset the thing.

I have cosseted my original large-flowered hybrids for nearly thirty years. I have faithfully cut the late-summer flowerers hard down every February, giving each a handful of sulphate of potash and covering their roots with the best compost to hand. I have given the big old plants a gallon of liquid manure – if not *quite* every week, then at least quite often through the flowering season. Two of them have rewarded me by becoming the most spectacular plants in the whole garden. The two are 'Victoria' and 'Perle d'Azur'. They are both trained out widely along walls, which is how the large-flowered hybrids look best. They have wires to climb along, and the young spring growth is attached to these with green twists until they submit to my plans and start hooking the steel-strong petioles of their leaves backwards into the places where I want them to go.

'Victoria's' wall is black – the tar-washed wall of an old clay-lump shed. But she has outgrown it, and usurped half the length of a neighbouring hornbeam hedge as well. From July to late August she is there, a 6m/20ft spread of big, perfect, bright heliotrope flowers, 14cm/nearly 6in across, brighter still in the freshly unfolding flowers where hints of cyclamen-pink lurk in the sepals' pleats.

But I have to make a deliberate effort to look at the detail of individual flowers. It is like standing too close to an oil painting when what you want to do is stand at ease and let the whole picture make its effortless effect. The difference between a single clematis flower and several hundred such flowers touching each other and overlapping in full bloom along a wall recalls the difference between a single delphinium floret and the soaring spikes of a whole delphinium plant. You can pick individual clematis flowers and float them in a bowl and marvel at their size and hue, but for the total clematis experience you have to be overwhelmed by continuous repetition in spreading space.

'Perle d'Azur' is my favourite clematis of all – perhaps because it is the one that seems most at home here. It is all a matter of getting a plant into the right place: it happens, in the end, by chance. My 'Perle d'Azur' must have found something particularly moist and rich for its roots to feed on beyond all my preparation; it faces east, against a whitewashed wall, and has the option of climbing right round a corner to the south. But it is not as hungry for space as 'Victoria', though just as long-lasting and floriferous; and it is blue. Clematis blue is notoriously difficult to reproduce in a photograph; most film seems over-sensitive to the violet that lies buried below the blue surface in a clematis flower. But 'Perle d'Azur' is catmint-blue, with six rounded sepals on long stalks that present the flowers to you at the most complimentary angle: the flowers are held vertically, you observe them face to face, they jostle for a place in the front row, they are a crowd. And between the open flowers the pointed buds of chartreuse-green (the same colour as the stamens) promise that the parade will continue for at least six weeks. For a few of those weeks *Buddleja* 'Lochinch' sends out its tapering cones almost to touch the clematis flowers, and a tub of agapanthus sends a related blue up into the air. These are coincidences. Farther along the wall there is a planned effect: a pink clematis grows and, in a good year, touches 'Perle d'Azur'.

It is 'Comtesse de Bouchaud', a little less vigorous and a little earlier to open (late June to late July). It plays a lovely supporting role; if it does not make it to my Top Two, it is certainly in my Top Three. It is a sort of lilac-pink; like 'Perle d'Azur', its flowers 8cm/3½in wide, made up of six sepals, each with four marks down the centre as if pressed by a fork into soft pastry. The sepals reflex backwards, round rather than pointy; the stamens are creamy-green. The flowers are not carried parallel to the wall, but have an upward turn, and all too soon the show is over, and nothing is left but seed-heads at their end of the wall, while at the other 'Perle d'Azur' is still a bank of blue.

Mixing two clematis together is good gardening sport, less fashionable nowadays than mixing clematis with pyracantha or chaenomeles or ceanothus, or with lilac or laburnum. In my experience, if the clematis is large-flowered, the shrubs tend to resent being deprived of light. Mixing clematis with climbing roses is another matter. My *C. flammula* is already lying lightly along the strong and glossy leaves of the 'New Dawn' rose, and just when the pale pink rose petals are falling the little white crosses of the clematis flowers are opening. Some people like to match colours together – to put the pink 'Hagley Hybrid', for instance, through the salmon-pink climbing rose 'Meg'; or, like me, to put *C.* 'Etoile Rose' through 'Mme Isaac Pereire'. In my case the matching is proving a mistake; the casual observer will not notice the small magenta clematis bells at all, clinging in the shadows of the huge, many-petalled magenta roses. Perhaps 'Etoile Rose' should share an arch with the small-flowered cream 'Albéric Barbier', or with 'Golden Showers', or perhaps take over, in summer, from the spring-flowering 'Frühlingsgold'.

I look back through my catalogues and begin to devise new schemes, and to hanker after new varieties. What about the artistocratic-sounding, blue-blooded 'Mrs Cholmondeley' to mix with 'Constance Spry', but far outdistance her length of blooming? What about 'Lady Betty Balfour', the latest of all the large-flowered hybrids to flower, in full violet flight in September and October if given a warm corner? What about the whole race of viticellas, to mix their nodding flowers with silver-leaved shrubs or herbaceous plants or heathers?

I have concocted a shortlist of the clematis I wish I grew, and I have arranged them in chronological order of their flowering, from early spring to the approach of winter. They will serve to sum up my gardening year among the family Ranunculaceae.

If my garden was in London, or in any reasonably warm and sheltered place, I would start with *Clematis armandii*, so strong, so handsome, prepared to cover whole trellises and fences with its glossy, evergreen, deep-lobed leaves and sprays of scented cream flowers in March.

ABOVE: *Clematis* 'Perle d'Azur' is the bluest clematis of all.

RIGHT: *Clematis × durandii*, the dark blue, herbaceous clematis, cannot climb but props itself over other plants and flowers for weeks.

I would follow it in April with *C. alpina* 'Frances Rivis', whose anemone-blue flowers have the four pointed, drooping sepals of all the alpinas, but are particularly big and beautiful. She could be fitted in anywhere, perhaps over low walls or rocks to remind her of her alpine origins; she does not need much space or sun.

Nor does my next spring flowerer: *C. macropetala*; it can grow in a big tub or pot. It is semi-double; its outer sepals are over an inch long, smoky lavender-blue; its inner rings are smoky lavender-grey, its central nectaries are cream, its stamens are pale yellow; it is an exquisite gradation of colour from deep to pale in staggered layers of pointed ruffles like a jester's collar. It is said not to need pruning, but it needs something that I have not given it; twice it has died in my garden, but it *should* be a flurry of flowers through April into May.

In May I would like to add a wisteria-blue clematis: 'Alice Fisk' – a cross between the blue but floppy 'Mrs Cholmondeley' and the shapely but rather common mauve 'Lasurstern'; she takes the best from both; she spreads 'Mrs Cholmondeley's' superior colour over 'Lasurstern's' superior shape ('Mrs Cholmondeley's' sepals being rather pinched at the base where 'Lasurstern's' are plump and crenulated).

In June I would welcome white as a rest from pink roses, but it should be a showy white of 'Jackmanii Alba', opening huge and double for its first flowering with a single-flowered repeat in September.

For an even showier white (or cream), I would try the exotic *C. florida bicolor* (syn. *C.f.* 'Sieboldii') if I had a sufficiently warm spot for it and felt in a sufficiently sophisticated mood. Unlike most clematis, this is a plant where the single flower is more eye-catching than the mass effect: rich cream sepals surround a central boss of purple, petaloid stamens with pale green carpels in the middle; it is as striking as the centre of a passion flower.

From June to September I would enjoy the darkest clematis of all: 'Niobe'. The colour is like the dregs of port wine – six thrilling, deep velvet, pointed sepals round stamens of lime green.

July would see the flowering of *C. × durandii* in

a mixed border. It is herbaceous, a cross between *C. integrifolia* and *C.* × *jackmanii*. Its large leaves do not know how to climb so it has to be tied to something if it is not to collapse on the ground. But its big flowers must not be allowed to hide at soil level; they must be lifted up and marvelled at. They are a unique indigo, almost navy blue, on four deep-veined sepals with cream stamens. They leave my gentle herbaceous clematis *davidiana* standing at the post.

For September and October I would choose another herbaceous clematis: *C.* × *jouiniana*, not because of its colour (palest milky-blue), nor because of its flower shape or size (the flowers are small and tubular) but because of its overall abundance. The flowers come in trusses; the trusses come from every leaf-axil; it is a mound of flowers, one of those plants that make people stop and ask: 'What's that?' as they pass it in the September border. The leaves are quite large, three-lobed, fresh green. It can be spread over the old, tired stems of other plants or tied up to a hedge of spring-flowering *Rosa* 'Canary Bird', or left as ground-cover, a mist of blue at the feet of pink Japanese anemones.

In November and December I would choose the native *Clematis vitalba*. I have seen it from train windows, journeying through Kent between Victoria Station and the coast, and it has looked like festoons of white roses, or like snow along the hedgerows. These are the seed-heads that give it the pet-name of old man's beard; its white flowers in June are insignificant by comparison. It has grown in Britain since the sixteenth century, and Gerard saw it here:

> in the borders of fields among thornes and briers, almost in every hedge as you go from Gravesend to Canterbury in Kent . . .

It was he who gave it its other name of Traveller's Joy, because of its habit of 'decking and adorning waies and hedges where people travel . . .'

I would be happy to stand quite still and see through a window the Traveller's Joy hooking itself to the hedges at the bottom of the garden as the year came to its close.

EPILOGUE

The Findings

Rana means 'frog'; *ranunculus* means 'little frog'. Perhaps the buttercup got its botanical name because it leap-frogs along the ground, but more likely it was because of its amphibious habits. The name suits the family. It is not only the clematis that thrives when watered; so do the delphinium, the kingcup, the monkshood, the Japanese anemone – never mind the marsh marigold. A few members of the family can turn on a decent show in dry conditions, but all of them rejoice in watering.

They are not found in tropical or subtropical regions. They belong to temperate lands, deciduous woodlands and rich pastures. They don't much like peat; most of them prefer an alkaline to an acid soil and many of them thrive in shade.

They are a poisonous family. 'Bane' recurs in their common names: the aconite is the 'wolfsbane', the cimicifuga is the 'bugbane', the actaea is the 'baneberry'. There are alkaloids present in their tissues, and the Victorian horticulturalist, Mrs Loudon, claimed to find a thin, acrid, yellow juice when she pressed their leaves or stems. I hoped this would make them rabbit-proof, and fancy that it may in mature plants, but tender seedlings of aquilegia, larkspur and love-in-a-mist prove irresistibly tempting to my rabbit population.

They are not noted for scent, though a few species clematis smell of almond or vanilla essence, the heavy, honey scent of *Cimicifuga ramosa* 'Atropurpurea' wraps round you as you pass, *Anemone sylvestris* has a gentle, woodland smell, and *Helleborus foetidus* stinks.

I embarked upon my year among the Ranunculaceae mainly because all the genera grouped under this name seemed so oddly different from each other. Now, at the end of the year (and for the rest of my life) I have lost that sense of oddity and unquestioningly accept the botanical family based on Linnaeus's systematic classifications in the eighteenth century. Linnaeus examined the stamens and pistils of flowers and grouped plants according to his findings. He found that in certain plants the stamens were numerous, and arranged in a circle, bending freely outwards above a circle of free sepals (and sometimes petals); they were attached to the base of a convex receptacle, and above them were numerous free carpels (ovaries) spirally arranged; when the seed was fertilized the stamens dropped, the carpels enlarged. Plants which shared this arrangement of parts he grouped together and labelled the 'Ranunculaceae'.

All year, my house has had seed-heads of Ranunculaceae ripening in jars. The carpels have been composed, typically, of five follicles arranged together in an urn shape, green and shining at first and finished at the tips with thin, sharp points. Winter aconites, marsh marigolds, columbines, thalictrums, delphiniums, monkshood – all in turn have presented the same sharp profiles; only in love-in-a-mist have the follicles spread wide and flat. I read that nearly all Ranunculaceae carpels contain a single seed (although in the actaea the seed is a berry); but the hellebore has two seeds to a carpel, which recently prompted a taxonomist to suggest expelling it from the Ranunculaceae and setting it up alone under the title 'Helleboraceae'. I am glad this idea did not catch on. But I am a gardener, not a botanist. I found it difficult to sustain my interest in jam jars of ripening seed-heads when, outside in the garden, the next batch of 'buttercups' was coming into flower.

Without doubt it was the discovery of coloured sepals (rather than petals) in flower after flower that struck me most. For the first five months of the year the theme was simple: a ring of coloured sepals (from five to sixteen) round a rich boss of stamens, often with honey-petals – that is, petals with a small honey-secreting area at their base or apex – in between. But from May onwards the variations were setting in: the honey-petals grew fly-away spurs, or the sepals grew into hoods, or the stamens metamorphosed into petaloid shapes, producing double flowers.

When beautiful sepals are shut, they make beautiful buds. From the winter aconite to the cimicifuga, the buds were often perfect globes – 'blobs', as the marsh-marigold fancier would call them, tiny blobs in the thalictrum, sizeable ones in *Clematis montana*. But when the open flower is a star, not a rounded daisy shape, then the buds are long and elegantly pointed (as in long-spurred aquilegias and large-flowered clematis). The delphiniums and monkshoods are the odd men out – neither globular nor pointed; their oddities of structure are prefigured in the dolphin- and helmet-shaped profiles of their buds.

The tall genera come in the second half of the year – thalictrums, monkshood, cimicifugas, Japanese anemones; they have in common exceptionally strong stems which can stand in the wind without props. Only the hollow-stemmed delphinium needs staking.

All the spring flowers and several of the summer ones grow from tuberous or rhizomatous roots, and many of them have the excellent habit of forming clumps: winter aconites, adonis flowers, hellebores, anemones, marsh marigolds, ranunculus, trollius, delphiniums, monkshoods, cimicifugas are all clump-forming; only the aquilegia and clematis, the common buttercup and the annuals are not.

My early warning is confirmed: there is no single characteristic that is common to all the genera in a plant family. The creeping buttercup does *not* have coloured sepals; the delphinium does *not* have numerous stamens round a central boss; the larkspur *can* endure a degree of drought. Though the typical leaf of the family is compound, either three- or five-partite and sharply dissected and toothed, the marsh marigold and the celandine have simple rounded leaves. In almost every genus where leaves grow on the stems as well as from the base, those leaves are arranged alternately, *except* in the clematis, where they are arranged opposite each other, in pairs.

Yet, as I wrote the lists for the back of this book, I found the cultural instructions for genus after genus were ridiculously repetitive, the same words chiming like a child's nursery-rhyme: 'fertile loam', 'leaf-mould', 'well-drained', 'moist' . . . 'moist' . . . 'moist' . . .

I picture a cool, moist, well-drained, fertile garden in which to display the family. It is long and narrow and runs down a gentle slope to a ditch that never dries out. At the top is the house, with *Clematis montana sericea* wreathing the garden door; at the bottom is a group of two or three deciduous trees where *Clematis flammula* and *C. tangutica* climb. A garden shed is completely hidden by the trailing growths of *C. montana* 'Tetrarose'. On a terrace outside the house, *Clematis alpina* and *C. macropetala* trail out of tubs where

Celandines are weeds or beloved wild flowers, according to taste.

brilliant Asiatic ranunculus flower in May. A flagstone path leads down the length of the garden from the house to the trees; celandines grow in its cracks and hepaticas get their roots beneath its edges. On each side of the path is a long border; one side is sunnier than the other, and also narrower, for the path is not quite central. In the sun delphiniums grow, arranged as Miss Jekyll instructs with the darker ones in the distance, the paler blues near the house. White delphiniums – the Pacific Hybrid 'Galahad', perhaps, 'Swan Lake' or 'Snow White' – grow among the pale blues, and among the dark blues there are dusky pinks and lavenders. Larkspurs, with their pinks and blues and mulberries, are sown throughout this border, and one or two airy belladonna delphiniums find room at the front. But just as a surfeit of delphiniums is threatening they are all cut down, and the two herbaceous clematis, *C.* × *durandii* and *C.* × *jouiniana* are pulled over them and encouraged to fill the spaces where they have been. Earlier, thalictrums have diluted the delphinium dose: groups of the very tall *Thalictrum speciosissimum* open their puffs of lemon-yellow among the varied blue spikes along the back, and at the front the blue-green leaves of *Thalictrum aquilegiifolium* make rounded clumps along the flagstones. At the far end of the border, just short of the trees, the tall and airy *Thalictrum delavayi* is planted as a marker.

On the opposite side of the path grow the hellebores: the evergreens and the herbaceous sorts, the short and the tall, the dusky pinks and plums, the apple-greens and the whites. The glossy new leaves of Lenten roses form clumps along the border's edge; farther back rise the darker, leafy stems of *H. foetidus* and *H. argutifolius*. For the sake of symmetry, some plants of the sun-tolerant *H. argutifolius* are allowed towards the back of the delphinium border, too. If only peonies were still classed as Ranunculaceae, shining red stems of peonies would be pushing through among the hellebores. Instead, in May, the little white flowers of Fair Maids of France are dotted in the shadows. Then suddenly, from midsummer onwards, the shady border is crowded with tall plants (that is why it needed to be wider than the sunny bed); there is a procession of monkshood, Japanese anemones and cimicifugas, balancing the delphiniums in July and out-flowering them from August to October. Down at the border's end, and opposite the tall marker of *Thalictrum delavayi*, stands an even taller specimen: the soaring *Cimicifuga ramosa* 'Atropurpurea'. These two are herbaceous sentinels at the bottom of the path in summer.

In January and February, when the sentinels are sleeping, you can see through to a colony of winter aconites under the bare trees. A month later, *Anemone blanda* spreads in dappled sunshine and *Anemone nemorosa* in dappled shade. Later, still, when the trees are in full leaf and casting shade, actaeas make a ferny ground-cover beneath them.

Everything in this garden seems to spread and there is prodigal self-seeding. Love-in-a-mist is allowed to choose its own flowering positions; columbines are everywhere, on either side of the path, in sun and in shade, wherever there is space.

Along the ditch the marsh marigolds establish themselves, growing thickly between the trollius flowers along the banks.

As winter approaches, however, hibernation threatens. Autumn colouring is brief and partial: the leaves of *Thalictrum speciosissimum* turn clear yellow, the leaves of *Cimicifuga rubifolia* turn amber, some clematis manage a half-hearted toast-brown, and the bold berries of *Actaea alba* flaunt their white and rosy summer colour-scheme. But soon the annuals are dead; the herbaceous plants are underground. There are no woody shrubs in the Ranunculaceae family. Early in spring, a thick mulch of leaf mould, or mushroom- or garden-compost is spread all over the borders, uniform and rich. But what about December, January? It's all right: the adonis flowers are there (*Adonis amurensis*), planted very near the house where they are visible through the windows, balancing the yellow aconites at the bottom of the garden. On the far side of the ditch, wild swathes of old man's beard obscure the boundary in December. In the two borders, the evergreen hellebores stand in as shrubs. But most important of all in winter, the shining, tapering olive-green leaves of *Clematis armandii* cover a long length of trellis at the back of the sunny bed.

The garden is never without flowers. Usually there are several genera in bloom at once. In March, for example, when *Clematis armandii* is smothered in its sprays of starry flowers and *Anemone blanda* is opening beneath the trees, the hellebores are still in flower, their stamens have fallen but the sepals look like fixtures, shapely as ever, though tending now towards the green rather than the white or pink. Even the yellow globes of January's winter aconites are still visible, half-swamped in the green sea of their leaves. For a feature of the 'buttercups' is longevity.

There is a great deal of blue in this garden, and a great deal of yellow. And though almost all the leaves are lobed and multi-partite, there is every sort of leaf-surface from light-reflecting to matt to frosted, almost every sort of green, as well as plum (in *Clematis montana* 'Tetrarose') and glaucous blue (in the thalictrums) and variegated (in certain columbines). It should escape monotony.

The finishing touch of beauty comes on the fences; here the large-flowered clematis hybrids climb. They are trained right down the shady side of the garden, and the varieties that need sun (like 'Lady Betty Balfour') are planted behind the delphiniums, climbing through the stems of *Clematis armandii*. But wires are fixed all along the fences, and trellises are fixed above, and from March until October there is sure to be a French 'Madame' or English 'Mrs', a 'Comtesse' or a 'Lady', an 'Alice' or a 'William' or a classic 'Niobe' in flower.

Pink touches blue and lavender; white sets off jewel colours, amethyst and ruby. And suddenly I see, in my mind's eye, one final family likeness: the pink of 'Comtesse de Bouchaud' is the same pink as the Japanese anemone; the subtle red of 'Mme Edouard André' is the same red that is found in certain larkspurs; the bright heliotrope of 'Victoria' is there in *Thalictrum aquilegiifolium* 'Thundercloud'; the blue of 'Perle d'Azur' is there in the delphinium 'Alice Artindale' when she's full-blown; the dark blue of *C.* × *durandii* is echoed in the late monkshood, *Aconitum carmichaelii*, and the yellow lanterns of *C. tangutica* are like upside-down cups of yellow trollius. It is as if an artist with a distinctive palette had been at work, dipping into the same set of pigments to mix the colours in this imagined garden.

The trees, of course, must come from other families, as there are no trees within the Ranunculaceae. But the Magnoliaceae come near in the sequence of plant families, so perhaps two magnolias spread between the end of the flower borders and the ditch. Beneath them I envisage a rough patch of grass. Here, at Easter time, the Pasque flowers rise; in summer, the wild pheasant's-eye adonis opens its blood-red flowers. And here the common buttercup creeps along the ground, a reminder that the most beautiful of plant families takes its common name from a weed.

CULTIVATION NOTES

ACONITUM (ak-o-*ni*-tum)

Monkshood

Monkshoods are so easy to grow that they invite neglect. But the more they spread, the poorer their individual flower-stems become, so every few years they should be lifted, divided, and the best young pieces replanted. They thrive in shade as well as sun, but appreciate a mulch in spring since, like all their relations, they are moisture-lovers.

A. carmichaelii: This tall, strong, late-flowering species (September–October) used to be called *A. fischeri*. Its leaves are sharply incised and particularly handsome, and its flowers are sometimes described as light Wedgewood-blue, but there are several varieties, of which 'Arendsii' and 'Kelmscott' both have the usual dark blue monkshood flowers.

***A. napellus*, the common monkshood**: many garden hybrids come from this. Among its offspring are the following named varieties:

***A.* 'Bressingham Spire'**: ramrod-straight, sapphire blue, up to 90cm/3ft tall, with glossy, fine-cut foliage on dense spikes; July–August.

***A.* 'Bicolor'**: slightly taller than 'Bressingham Spire', 105cm/3½ft with flowers on which the blue seems smudged over white, and the flower-stems branch sideways as well as upwards; July–August.

***A.* 'Ivorine'**: quite distinct, small 75cm/2½ft with creamy-ivory flowers which come first in late spring, and again in autumn.

***A.* 'Newry Blue'**: an early-flowerer, making great patches of deep blue in the June border. It is a spreader, inclined to be invasive, but hard to say 'no' to.

***A.* 'Spark's Variety'**: the tallest of all, 150cm/5ft sometimes blown aslant in a wind, but superbly aristocratic with its poised, wide-branching stems and the aquiline profiles of its indigo-blue flowers, each one aloof from its neighbours. July–August.

A. vulparia (A. lycoctonum): like 'Ivorine', this is a pale creamy yellow, but 120cm/4ft tall, with branching stems that may need staking, unless it can be placed so that it leans through and rests upon other plants.

ACTAEA (ak-*te*-a)

Baneberry

Woodland plants with poisonous berries, easy in moist shade, particularly on alkaline soil. Their flowers are inconspicuous; it is for their ferny leaves and berries that they are grown. They look at home among shrubs, where their rhizomatous roots can spread. Mulch with leaf-mould in the spring.

***A. pachypoda* (syn. *A. alba*), dolly's eyes**: fluffy white flower-heads at midsummer are followed by small white berries, each with a black dot, set on swollen pink pedicels and standing well above the ferny foliage, 90cm/3ft tall.

A. rubra: shorter (60cm/2ft), with clusters of incandescent, strawberry-red berries among the leaves, which could be mistaken for an astilbe's. Spreads vigorously and has the knack of looking indigenous in a shady border.

Aconitum napellus has five deep blue sepals, the upmost one being large and hollow – a hood to protect the nectaries and stamens within.

ADONIS (ad-*o*-nis)

Like its relative, the winter aconite, the adonis likes leaf-mould and a spot under deciduous trees, where the sun can touch its buds. Feed with bonemeal or fish-meal in autumn or early spring when its root system is growing. The perennials become dormant after flowering, so you should mark their positions with a stick.

***A. aestivalis*, pheasant's eye**: the annual summer wildflower with small, blood-red, single flowers and feathery leaves, 25cm/10in tall. Easy to grow from seed sown where it is to flower.

***A. amurensis* 'Fukujukai'**: a winter perennial, one of the very first to open – January–March. Its deeply divided leaves start bronze, but turn green, its yellow flowers are 5cm/2in across and grow on branching stems 15cm/6in tall.

A.a. 'Flore Pleno' (syn. *A.a.* Plena): a double version not unlike a marigold, yellow with a green centre, flowering in March–April.

A. vernalis: a flower of spring, with shining lemon-yellow buttercup flowers, 5cm/2in across, resting on feathery ruffs of leaves 25cm/10in tall.

ANEMONE (an-*em*-o-nee)

Wind flower

SPRING-FLOWERING ANEMONES

The early-flowering anemones are sold by bulb merchants in the autumn, but they are not bulbs; they are corms or tubers, small, brown and knobbly. You should soak them overnight before planting, and discard any that have not plumped up. Plant corms 5cm/2in deep with the convex side downwards, if discernible, in light, well-drained, preferably alkaline soil. Under deciduous trees is ideal for most of them – they need the touch of sunshine to open their flowers. When their fine-cut leaves first appear, spread a mulch round them of leaf-mould or garden- or mushroom-compost: this will set off the flowers as well as keeping the plants moist.

A. apennina: azure-blue, slightly cupped flowers, 3.5–4.5cm/1½–2in across, grow from a thick, tuberous rhizome and flower in March–April. Its seed-heads remain erect after flowering, which best distinguishes it from the very similar *A. blanda*. It spreads both by seed and by its horizontal rhizomes.

***A. blanda*, the Grecian windflower**: its 3.5–4.5cm/1½–2in flowers have many pointed, slender sepals, spreading into a daisy shape, typically blue, but also white or pink, and the seed-heads bend over. It grows from a knobbly tuber to a height of 15cm/6in, and should be planted in a warm, sheltered place to encourage its March flowers. If happy, it colonizes freely.

A.b. 'White Splendour' is a dazzling white variety with particularly large and long-lasting flowers.

***A. coronaria*, the poppy anemone**: flowers May–June on 25cm/10in stems in red, blue, violet, saffron-yellow. Plant in rich, sandy loam, but do not expect it to last for many springs.

A.c. de Caen Group: single flowers with prominent black anthers.
A.c. St Brigid Group: double flowers, many-layered and blowzy.

A. fulgens: bright scarlet, single flowers with black anthers above a deeply cut involucre (25cm/10in tall). Flowers March–May if given sun and good drainage.

***A. hepatica* (now *Hepatica nobilis*)**: beautiful but temperamental, a lime-loving, anemone-like plant 15cm/6in tall, with pale blue flowers and lustrous three-lobed leaves, blooming from February to April when happily established.

***A. japonica* (now *A. hupehensis* or *A.* × *hybrida*)**: see below.

***A. nemorosa*, the wood anemone**: the only native British anemone, its rhizomatous roots spread freely in semi-shade and even tolerate acid soil. It is easy, provided it is not allowed to dry out. The simple April flowers have six or seven white sepals, tinged with pink on the reverse, 2.5cm/1in across, 10cm/4in tall.

A.n. 'Robinsoniana': a pale wisteria-blue variety.
A.n. 'Vestal': tiny button flowers densely packed with petaloid stamens.

A. *pulsatilla* **(now *Pulsatilla vulgaris*), the Pasque flower**: a native of limestone country, this will not grow on acid soils. The big, single, jewel-coloured flowers (amethyst, ruby, pearl, with rich gold centres) open in March from silky, silvery involucres on 30cm/12in stems with fine cut leaves. A 'must' for chalk gardens where all it asks is sun.

A. *ranunculoides* **(now correctly *A.* × *lipsiensis*)**: a small yellow anemone, 15cm/6in tall, with a finely scissored involucre like a winter aconite's, it spreads in woodland by means of its horizontal rootstock, and flowers in March–May.

A. *sylvestris*, the snowdrop anemone: so-named because it hangs its head in bud. It opens satiny-white and scented, and flowers from April onwards, resembling a small white Japanese anemone, 30cm/12in tall. Given light, moist, loamy soil it will spread in sun or shade.

JAPANESE ANEMONES

Long known to gardeners as *Anemone japonica*, these autumn-flowering anemones are now variously classified as *Anemone hupehensis*, *A.h* var. *japonica* and *Anemone* × *hybrida*. They flower for three months, from August to October, need no staking or pampering, will perform in shade as well as sun, and spread steadily, filling the empty spaces of the late-summer garden. But don't expect much in the first year. Whether you buy a plant from a garden centre, or beg a piece from your friend's clump, it probably will not flower for at least a year, however well you water it (though, being a buttercup, it does like water). Its flowers shrink in very dry summers. I list a few hybrid varieties grouped by colour.

PINKS

'Bressingham Glow': a relatively short variety (60cm/2ft) for the front of the border, with profuse, rosy, semi-double flowers.

'Queen Charlotte': a much-favoured beauty with large, outward-facing, semi-double rose-pink flowers, 75cm/2½ft.

'September Charm': simple, single flowers on stems 75–90cm/2½–3ft tall. Despite its name, it starts to flower in August and continues until the frosts.

WHITES

'Honorine Jobert' (or *A. japonica* 'Alba'): the old-fashioned favourite, seen in gardens since 1858 and still unsurpassed, 120–150cm/4–5ft tall. Single white flowers with yellow stamens on slender, branching stems, blooming from August until October.

'Luise Uhink': an old German variety, 90–120cm/3–4ft tall, with semi-double white flowers.

'Whirlwind': hybridized in America in 1887, differs from 'Luise Uhink' in having slightly smaller, semi-double flowers which are quilled along the edges.

AQUILEGIA (ak-wil-*ee*-ji-a)

Columbine

Aquilegias are at home on chalk, but will thrive in any neutral soil provided it is well-drained, moist and fertile. They will grow in sun or shade, but are happiest in half-shade, with shelter from wind. You may still have to stake the tall varieties with twiggy sticks. Once established, don't try to move them or divide them – they have long tap-roots. They may seed themselves in your garden; certainly they are easily raised from seed if you want to grow numbers of them. Sow the seed outdoors in May, and move seedlings to their permanent positions in September–October. They are not long-lived (and are delicious to rabbits and mice, despite having seed that is poisonous to humans) but with luck and a mixture of strains you may extend their flowering period from May to July.

SPECIES

A. alpina: usually quite short (30cm/12in), this plant has delicate, nodding, bright blue flowers.

A.a. 'Hensol Harebell': true to its name, it is like a blue and white harebell, about 25cm/10in tall.

A. canadensis: a North American species with glowing flowers, red-spurred and lemon-yellow centred, 60cm/2ft.

Aquilegias look loveliest all mixed together in their pastel colours: Grannies' bonnets in pink, purple, white and rose.

A. formosa: another North American species with small, fly-away, scarlet and gold flowers on tall, branching stems, 75cm/2½ft.

A. longissima: the flowers are almost white with very long spurs; a species originating from Texas to Mexico, it is the important ancestor of the Californian long-spurred hybrids. 60–90cm/2–3ft tall.

***A. vulgaris*, the 'columbine' of Elizabethan England**: in its double forms, it is called granny's bonnet. The flowers may be blue, pink, purple or white on 30–60cm/1–2ft stems, and the spurs are incurved and strongly hooked. A prolific self-seeder.

A.v. clematiflora (syn. *A.v. stellata*): this 'clematis-flowered' aquilegia, developed in the 1930s, was so called because it is spurless, and the five starry, spreading sepals recall certain clematis flowers. There is a white form, *alba*, and a pink, *rosea*, as well as blue. A prolific self-seeder.

A.v. 'Nivea' (syn. 'Munstead White'): this choice variety was renamed because it was much-favoured by Miss Jekyll; it has large white flowers offset by emerald foliage frosted with grey on strong, vigorous stems.

A.v. 'Nora Barlow': a sturdy old double variety, once called the 'rose columbine'. The spurless flowers are fat pincushions of creamy green and pinkish red. Tall (60–90cm/2–3ft), strong and in flower from May to July.

LONG-SPURRED HYBRIDS

Two famous strains were raised in California a hundred years ago: McKana Giant Hybrids, and Mrs Scott-Elliot's Strain. They both have strong cutting stems, 75–90cm/2½–3ft tall, wiry and branching, and the elegant flowers with their straight, slender spurs include red and yellow as well as blue, pink, violet and white. They bloom in May and June. Ancestors of modern F_1 hybrids.

A. 'Crimson Star': a striking, long-spurred variety with raspberry-red sepals and white petals (45–60cm/18–24in tall).

A. 'Dragonfly': a lovely seed-mixture with pastel colours of violet, lilac, pink, primrose, white (45cm/18in tall).

CALTHA (*kal*-tha)

Marsh Marigold

A plant for wet places – a bog garden or the shallows of a garden pond. If you can give it continuous moisture at the roots, it will be no trouble.

***C. palustris*, the common kingcup**: a wildflower flourishing in marshes and along streams. It creeps and roots along damp verges, growing 45cm/18in tall, with bright green, heart-shaped leaves below pure yellow single flowers 5cm/2in wide. The flower-stems branch, each holding aloft its array of round buds and open cups in April. Effortlessly rewarding beside a big garden pond.

C.p. alba: a compact variety, 15cm/6in tall, with waxy white flowers on prettily branching pinkish stems and a long flowering season, from March till May. If you have no boggy soil round your garden pond, you can grow it in the pond in a pot raised on bricks so that the crown of the plant is set just above the water.

C.p. 'Plena' (syn. *C.p.* 'Flore Pleno'): prolific orange-yellow flowers in tight, many-layered rosettes among bright green rounded leaves on a 30cm/12in plant. If you like the colour and the style, it is a stalwart 'doer', and supplies a bright April show.

***C. polypetala* (now correctly *C. palustris palustris*)**: an altogether larger plant than *C. palustris*, this can grow 60cm/2ft tall, and spreads by stolons or rooting stems widely over the water, its leaves as large as an arum lily's, its single yellow cups larger, too, than *C. palustris*, but also less prodigal, more occasional, and a little later (April/May). It is happy planted 15cm/6in below the water.

CIMICIFUGA (sim-is-if-*u*-ga)

Bugbane

Cimicifugas like shade. Their leaves scorch in hot sun (only the thick leaves of *C. cordifolia* are exempt from this). Give them a cool, moist spot with good drainage, and mulch them in spring. Though tall, they need no staking. They return year after year, neither spreading nor diminishing, but behaving beautifully, just when most favourite flowers are over.

C. racemosa: somewhat softer-looking in leaf and flower than *C. rubifolia* (the long, white racemes bend over in graceful curves); also fussier – it needs moisture and shade to do its best. Flowers August–September, 150cm/5ft tall.

***C. ramosa* 'Atropurpurea'** (now *C. Simplex* Atropurpurea Group): a star performer. It takes the whole season to complete its act; first come finely cut, dull purple leaves, then the stiff, strong purple stems start to grow. They reach 180cm/6ft before the purple bead-buds break in September–October to reveal pinkish-cream stamens. It does not enjoy being in a crowd, and will adorn a prize site in the garden standing alone – at the end of a border or the turning of a path.

***C. rubifolia* (syn. *C. racemosa cordifolia*)**: a handsome North American plant whose leaves are like the Japanese anemone's, but paler green. Very tall (150cm/5ft), erect and strong, flowering in August when its tight, brownish buds open to reveal bottlebrushes of cream stamens. Rare, but trouble-free.

C. simplex: slightly shorter (90-120cm/3-4ft) and later to flower (October), this is a precious acquisition in the autumn garden:

C.s. 'Elstead': opens pure white, but its buds are purplish and its stamens pink.

C.s. 'White Pearl': ivory, tinted pale green.

CLEMATIS (*klem*-a tis)

Your new clematis will be pot-grown; you may plant it at any time, but choose the site with care: clematis dislike being moved. Make sure the planting hole is deep (at least 45cm/18in), enriched with humus, and that the roots will be in the shade; if necessary, arrange a slab of stone or granite chippings over them. Place the stem so that at least one pair of leaves comes below ground level. Slant the cane slightly towards the tree, wall or arch up which the plant will climb. Before you infill the earth, it is an idea to set an empty flowerpot beside the stem, covered with a tile, and you can feed liquid manure directly into this every week or two throughout the summer. All clematis need lots of water and food. When growth starts in February, sprinkle a good handful of sulphate of potash round the stem and spread a mulch on top.

Pruning: All the large-flowered hybrids that bloom on the new wood from June onwards are hard-pruned in late February, cutting down to the first pair of leaves on the previous season's growth. It is particularly important to prune a new plant severely in its first spring, to 15cm/6in above ground. With an old plant, you throw away armfuls of last year's tangled stems. If you don't, the plant will be a mess and the flowers will be out of sight. But the large-flowered hybrids that bloom in every summer on last year's wood need gentler pruning in later February: not to the first pair of leaf-buds but to the first *fat* pair of leaf-buds – a very different matter. Most of the small-flowered species need not be pruned at all until they have overflowed their allotted space, when they should be cut back after flowering. Each variety's preferred pruning is marked by an initial letter at the end of its entry.

H = hard pruning in late February.
L = light pruning in late February.
N = no pruning unless plant is overgrown.

There are well over four hundred varieties of clematis available commercially; the list below is a highly selective personal choice.

LARGE-FLOWERED HYBRIDS

To simplify choice, they are grouped by colour. The flowers are all roughly 10–15cm/4–6in across unless described as 'large'.

BLUES

'Alice Fisk': a cross between 'Mrs Cholmondeley' and 'Lasurstern', combining the former's colour (wisteria-blue) with the latter's elegant shape (long, pointed sepals forming huge flowers); modest in growth (3m/10ft) and repeat-flowering, May–June, September. L.

C. × *durandii* (a cross between *C.* × *jackmanii* and *C. integrifolia*): a herbaceous clematis, that must be tied to its supports as its leaves don't cling. Superb in a small town garden, trained over shrubs that have finished flowering. Can spread for 2m/6ft along a border, with thrillingly intense, indigo flowers, fading to lustrous parma violet; 4–6 wide-spaced sepals distinctly ridged down their centres; June-September. A winner, to be ordered from specialist growers. H.

'Mrs Cholmondeley' (pronounced Chumley): the most generous of clematis – begins to bloom in May and has fresh waves of flower almost non-stop until September. The large flowers have chocolate stamens with lavender-blue sepals pinched at the base so that you can see space between them; some people call it 'blowzy' or 'floppy'; but it will grow to 4–5m/13–16ft on a north wall, and looks lovely on an arch mixed with a yellow rose like 'Golden Showers'. L.

'Perle d'Azur': as nearly blue as any clematis can be. The flowers are set on long, graceful stems, the sepals are blunted and recurved, the stamens greenish. It will make a wall of blue, 3m/10ft long, for six weeks in July and August. Easy, floriferous, irresistible. H.

MAUVES

'Victoria': rich heliotrope-mauve flowers with six sepals and silvery-green stamens. Exceptional performer, vigorous and blight-resistant, covering 4–5cm/13–16ft on any aspect, and flowering continuously from July to September. H.

'William Kennett': a hundred years old and still a top favourite, because of its very large and shapely lavender-mauve flowers, whose eight wide sepals overlap one another and have crimped edges; the dark reddish-brown stamens are like a rosette. Will grow 6m/20ft on a south or west wall. Flowers on the old wood, June–July, so fits well on a shrub like pyracantha, slotting in between that shrub's flowers and berries. L.

PINKS

'Comtesse de Bouchaud': keeps within bounds (3m/10ft). A mass of soft lilac-pink flowers with six rounded sepals and cream stamens in June–July. Mixes well with 'Perle d'Azur' or 'Madame Edouard André'. H.

'Hagley Hybrid': another neat, well-behaved clematis, seldom taller than 2.5m/8ft, the shell-pink flowers fading to washed-out mauve with six pointed sepals and brown stamens. Fully earns its keep in small town gardens, flowering abundantly in any aspect from June to September (but fades less in shade). H.

PURPLES

'Lady Betty Balfour': its special claim to inclusion is its lateness in flower (September–October); it is something to look forward to when all the other large-flowered hybrids are over. Its flowers have violet sepals and yellow stamens; they need a sunny wall for best performance (4–5m/13–16ft tall). H.

'The President': famous royal purple with a silvery stripe down the back of each pointed sepal, and deep reddish-purple stamens. First flowers in June on the old wood, when flowers are large. Later come smaller flowers on new wood in several bursts until September. A manageable height (3m/10ft) and thrives in sun. L.

REDS

'Madame Edouard André': the most persistent flowerer of all the hard-pruned clematis. It is not tall (3m/10ft); its energy goes into producing a succession of dusky mauve-red flowers for three months, from June–August. H.

'Niobe': so dark it is almost black, glamorous and striking with six pointed sepals and pale green stamens. (3m/10ft tall). H.

WHITES

'Jackmanii Alba': eye-catching flowers with milky-white sepals, very large and double in June, followed in later summer by a show of single white flowers on the new wood (3m/10ft tall). L.

'Marie Boisselot' (formerly 'Madame le Coultre'): large, shapely flowers of shining white and cream stamens, held with an upward tilt on a vigorous plant (5m/16ft). The broad, round sepals overlap, which intensifies the whiteness of the flower. The most popular of white clematis. L.

SPECIES AND SMALL-FLOWERED VARIETIES

***C. alpina* 'Frances Rivis'**: the largest-flowered of the *alpina* group, with masses of dangling, lavender-blue flowers in April/May; long, pointed sepals, white stamens (2–3m/6–10ft high). Train over a shrub or on a fence, preferably in a cool aspect. N.

C. armandii: a vigorous evergreen climber with glossy, leathery, three-lobed leaves and masses of small white scented flower-clusters in March. Handsome all the year round and ready to romp along walls and trellises, it will quickly furnish a new town garden, provided it faces south or west. Tender in exposed positions. (Height 4–7m/13–23ft.) N.

C. cirrhosa: an evergreen winter-flowerer for conservatory or warm and sheltered corner. Small, greenish-cream flowers dangle in January among glossy, ferny, three-lobed leaves. (Height 4m/14ft.) N.

***C.* 'Etoile Rose' (possibly a cross between *C. texensis* and *C. patens*)**: Americans call *C. texensis* the 'leather flower' because of the thickness of its sepals. The flowers, blooming from July to October, are crisply defined cerise Turk's caps with silver margins, better grown over an arch and so seen from below. (Height 2–2½m/6–8ft.) H.

***C. flammula*, the fragrant Virgin's Bower**: tiny white flowers in August/September, with vanilla scent which hangs on the air; silver seed-heads, delicate, ferny leaves on a vigorous, billowing plant. (Height 3–4m/10–13ft.) H.

***C. florida* 'Sieboldii' (*C.f. bicolor*)**: a uniquely striking clematis could be mistaken for a passion flower because of the green and purple boss of petaloid stamens in the centre of the large, greenish-white flowers. Early growth can be cut by frost, and it is far from vigorous (2m/6ft at best). A collector's piece for a warm spot in the June–July garden. N.

***C. × jouiniana* 'Praecox'**: semi-herbaceous, needs tying to supports unless allowed to sprawl as ground-cover or spread over tree stumps. Big panicles of little ash-pale, four-sepalled flowers which cover the whole plant to form a great mound of flowers (3–4m/10–13ft) in August–September/October. Best with some sun. L.

C. macropetala: dangling lavender flowers with long, pointed sepals like those of *C. alpina*, 8cm/3in across, but with a ring of pale, petaloid stamens inside giving a double effect both delicate and showy. Not vigorous (2–2.5m/6–8ft); sometimes planted to trail over low walls or even out of large tubs. N.

***C. montana* 'Tetrarose'**: a beautiful form with the largest flowers of the *montana* group; the four sepals are broad and satiny, rose-mauve, the stamens straw-coloured, the serrated leaves purple-green. It will cover large spaces, spreading 6–9m/20–30ft even on a north wall, and cover itself with flowers in May. N.

***C. spooneri* (syn. *C. chrysocoma sericea*; now *C. montana sericea*)**: very like a *C. montana* but not quite as rampant. Smothered in large white flowers in May/June. Performs well on a north wall. (Height 8m/25ft.) N.

C. tangutica: bright yellow nodding bell-shaped flowers with brown stamens produced on new wood in August/September, and followed by puffs of silver seed-heads among fresh green ferny leaves. Long-flowering – for two months it is a mass of flowers and seed-heads, wild-looking and vigorous, and particularly at home climbing through a tree. (Height 4–7m/13–23ft.) H.

CONSOLIDA (kon-*sol*-i-da)

Larkspur

The annual delphinium. Sow seed where it is to flower; plants do not move well. It is easy, but takes time to germinate, and likes sun and enriched soil. Scatter seed along shallow drills in spring for July–August flowers, or in early October for large June flowers, and thin out seedlings to 22cm/9in. Your seed packet will warn you that the seeds are poisonous. It will probably produce mixed colours: dark blue, pale blue, red, pink and white, but the flowers will be either double or single according to the variety and the stems either tall or quite short.

'Giant Imperial': spikes of double flowers, freely branching, 90–120cm/3–4ft. Grown commercially for cutting.

'Hyacinth Flowered': dwarf and compact, 30cm/12in.

'Rocket': spikes of single flowers with shapely spurs. The 'tall' rocket reaches 75cm/2½ft, the 'dwarf' 40cm/15in.

'Stock Flowered': fat spikes of double flowers on long stems, 90cm/3ft. Another florists' larkspur, grown for picking.

DELPHINIUM (del-*fin*-i-um)

Like the clematis, the delphinium should be planted where it is to remain; it does not like a move and does not divide well. Again like clematis, it is a hungry feeder; it likes lots of water, applied with a sprinkler or the fine rose of a watering can; it should never be allowed to dry out while the flower-spikes are forming. At this time it should also be fed with blood, fish and bonemeal, or a well-balanced liquid fertilizer. A new plant should be set with its crown 2.5cm/1in below the surface in spring; it may be thin and inconsequential in its first year, but from the second year onwards it will strengthen and give a major performance, provided you not only feed it well, but stake it and protect it from slugs. Spread sifted cinders, sand or bark round the crown in winter as slug protection. Push three or four canes or galvanized steel rods round the crown when the stems are about 30cm/12in tall; you need not tie individual stems to stakes – simply encircle the whole plant with a string, running from stake to stake. Do this twice in the growing season. (Or use tall, twiggy pea-sticks, or a patent galvanized steel frame for the stems to grow through.) When they have finished flowering, by late July, cut them to within 5–10cm/2–4in of the ground, water and feed them again; new growth will appear promptly and they may mount a repeat, autumn performance. When this is over, dead-head them but don't cut the stems down until the leaves have died; then cut them to ground level – you don't want water to collect in the hollow stems. When they are really established, they may become too thick. You should not allow more than ten stems per plant; eliminate the feebler ones. You can now take cuttings with a sharp knife where the stem joins the crown, making sure you get a bit of basal tissue attached. Hold cuttings under water, then stand them in a jar of tapwater on a window-sill. When the 'bush' of roots reaches about 1–2cm/½–1in, plant the cuttings in 10cm/4in pots of moist compost and put a split cane in now, rather than disturb the roots later. Plant out in the border in April–May and they may be in flower in late summer, when the mature plants are over.

Alternatively, and most economically, you can grow delphiniums from seed. Sow 1cm/½in deep in a spare part of the garden from April to August. The resultant plants will be a lottery; move your favourites, while still young, to prize places and discard the rest.

If you have a small garden with narrow borders, it seems wise to go for the shorter sorts – the Belladonnas or *D. tatsienense* – that need minimal staking. Tall delphiniums planted close to a wall lean forwards toward the light, rest their weight against the restraining string, and may break over in a high wind.

Belladonna delphiniums: relatively short (not more than 100cm/40in), the flowers are elegant, long-spurred and single, loosely spaced on wiry stems, and the plant branches generously so that the flowering season is prolonged from July to August, particularly with regular dead-heading. Famous old named varieties are:

'Blue Bees': pale blue with a white eye, earliest to bloom.
'Lamartine': Oxford-blue.
D. moerheimii: the original belladonna; pure white.
'Pink Sensation': introduced by the Royal Moerheim Nurseries in 1936, it was a natural pink mutation from a red species, *D. nudicaule*.

D. elatum: this species is the chief parent of the many hundreds of named cultivars that have been coming and going in nurserymen's lists since 1850. If you want a particular named variety, you will probably have to order from a specialist grower, but in a book about the Ranunculaceae family, I name just one: the so-called 'ranunculus-flowered' 'Alice Artindale'. Very tall (1.80m/6ft), with perfect deportment and tapering spikes of azure-blue tinged with lavender. The florets are fully double rosettes; the central 'bee' of the usual delphinium is metamorphosed into reflexing petals and the florets are well spaced up the flower spikes like rows of small ranunculus. A good cut flower, it also dries well – the petals don't drop until the whole flowering is over.

D. tatsienense: a short (30–45cm/12–18in), and short-lived, Chinese species which is sometimes used as an annual at the front of the border, and can self-seed if slugs are no problem. Intense, cornflower-blue flowers in branching panicles – airy and dainty.

GOOD HYBRID SEED MIXTURES

Blue Fountains: large, semi-double florets forming showy, tapering spikes (75cm/2½ft) on sturdy stems – blue, mauve, purple, occasionally white.

Connecticut Yankees: raised by Edward Steichen in USA, a nice selection of easily managed, branching (75cm/2½ft) plants with single flowers in pastel blues and mauves and whites. Not long-lived, but very prodigal of flower, mounting repeat performances, and blooming in summer from a March sowing.

Pacific Hybrids: these are the tall, spectacular beauties first bred in California by Frank Reinelt, who crossed *D. elatum* with *D. cardinale* and produced this race of pastel-coloured flowers, including pinks. They were originally more suited to California than to Britain, where they tended to be annual or biennial, but they have been further hybridized with hardier strains. They flower early and come true from seed (1.5–2m/5–7ft).

ERANTHIS (e-*ran*-this)

Winter aconite

Best planted 'in green', that is, with the leaves attached in spring. A clump taken from a friend's colony between February and April and swiftly replanted is the ideal. The small dry tubers are much harder to establish, but if they are all you can find, buy them as early as possible – in late summer – and cover the small, brown, spiky tubers with damp peat for a day or two before planting 5cm/2in deep in any well-drained soil. They prosper in leaf-mould under deciduous trees where the winter sun can coax the round yellow buds to open as early as January.

E. cilicica: a species distinguished by its very deep-cut, bronze-tinted leaves and its later flowering season (February–March).

E. hyemalis: the true winter aconite (*hyemalis* = 'pertaining to winter') flowering from January–March, with perfect yellow globes resting on bright green ruffs (involucres) 5cm/2in above the ground. Self-seeds freely.

HELLEBORUS
(hel-*le*-bor-us, commonly hel-le-*bor*-us)

Hellebore

Dig a deep hole for your hellebore and fill it with compost, but plant shallowly. Keep a new plant well watered in dry spells. Feed established plants with rose fertilizer or a general fertilizer like Growmore in early spring, and then mulch them with any well-rotted organic matter. Though they often seem to flourish untended, they are gross feeders and thrive in moist leaf-soil. They grow from poisonous rhizomes and gradually form large clumps; it is not easy to move them, nor is it easy to divide them or propagate them vegetatively, which is why named varieties are expensive. They seed freely, however, and unnamed seedlings are plentiful and rewarding. They make good evergreen ground-cover among shrubs, enjoy half-shade and tolerate full shade, and furnish the winter garden with rich leaves and subtly captivating flowers from January to March. One problem is leaf-spot disease which makes rusty-brown or black patches on the leaves; spray infected plants with a fungicide like 'Benlate' at fortnightly intervals when the buds are forming, and all old leaves should be removed routinely in winter from *H. niger* and *H. orientalis* to guard against infection.

***H. argutifolius* (formerly *H. corsicus*)**: this has a woody stem, 60–100cm/2–3ft, from which both leaves and flowers are produced. The leaves are dark green, three-lobed and saw-edged, and each stem is topped with a large head of apple-green flower-cups; an established plant has many stems. Cut out the old stems to ground level after flowering, to make way for the new. If you want to plant a hellebore in the sun, this is the one to choose. But it is easy anywhere, and a great self-seeder.

H. × sternii: intermediate hybrids between *H. argutifolius* and *H. lividus*, the green flowers can be flushed with maroon, or with pink, as in the Blackthorn Strain, which also has glaucous, marbled foliage. It does best in a sheltered spot, with some sun.

***H. atrorubens* (a name adopted commercially for a hybrid between the true species *H. atrorubens* and *H. orientalis*)**: its large, purple flowers, standing well above the foliage, are the earliest of all, often blooming before Christmas (30cm/12in tall).

***H. foetidus*, the stinking hellebore**: a British native of chalk downs; with dark green leaves which radiate from the woody stems like the spokes of an umbrella; the individual leaflets are pencil-thin, and the big clustered flower-heads of pale green bells stand above them. It has only one shortcoming: alone among hellebores, it has weak rhizomes and after a year or two many rot at ground level (60–90cm/2–3ft tall).

H.f. 'Miss Jekyll': a sweet-scented form.
H.f. 'Wester Flisk': distinguished by its red stems and flower stalks, and greyish-green leaves.

H. lividus: half-hardy, so best grown under glass, this has a pinkish-purple stain on the outside of the sepals and under the dark green, smooth-edged leaves (30–45cm/12–18in tall).

***H. niger*, the Christmas rose**: the big white flowers and deep green, spreading leaves grow on separate stout stems, and often the leaves half-hide the flowers. Not always easy; thrives in rich, rather heavy soil in partial shade. Cover the buds with cloches in January to encourage length of stem and protect from mud splashes (15–25cm/6–10in tall).

H.n. 'Potter's Wheel': a variety with large, rounded white sepals and a conspicuous green eye, strong stems, vigorous foliage. Feed with liquid fertilizer at fortnightly intervals in spring.

***H. orientalis*, the Lenten rose**: this starts to bloom in February and continues throughout March. The many unnamed hybrids have shaded, pendent flowers in cream, green, pink, maroon and smoky-purple standing on strong, separate stems well above the rich green pedate leaves. Taller than *H. niger* (40–60cm/16–24in) and easier, it enjoys moderate shade and moisture, and colours best on alkaline soils.

H.o. Ballard Strain: highly refined in their clear colours (from lime-yellow to inky blue-black) and in shape (they hold their heads high). These are costly collectors' pieces.
H.o. guttatus: a variant of *H. orientalis*, with red spots on the inside of its large, whitish flowers.

H. purpurascens: an early-flowering Balkan species (January–March), its flowers are deep purple on the outside, overlaid with grey, and greenish-purple within, on short, separate stems 20–30cm/8–12in tall.

H. torquatus: this variable species includes the very dark colour used by Mrs Helen Ballard to achieve her inky-blacks and purple-blues. The long-stemmed flowers open in February–March, and the leaflets are divided into many segments, hairy beneath (25–35cm/10–14in).

H.t. 'Pluto': an exciting combination of dusky prune colour on the outside of the flowers and pale jade-green inside. Needs a sheltered site.

NIGELLA (ni-*jel*-la)

An easy annual. Sow where it is to flower, just covering the seed with soil, and thinning seedlings to 22cm/9in apart. A March sowing will produce July/August flowers, but if you sow in September in a sheltered place, the flowers will be larger and will open earlier. They may need tactful propping with twigs unless they are growing where they can lean against other plants; they look good sown in blocks among old shrub roses. Deadhead to encourage longer flowering, but spare some of the decorative seed-pods in the hope of self-sown plants in unexpected places next year.

***N. damascena*, love-in-a-mist**:
N.d. 'Miss Jekyll': tall, sky-blue semi-double flowers 45cm/18in.
N.d. 'Persian Jewels': semi-double flowers in mixed colours – pale blue, purple, light pink, dark pink and white (45cm/18in).

***N. hispanica*, the fennel flower**: taller (60cm/2ft), this has larger, single, mid-blue sepals spreading out flatly round a central boss of prominent dark stamens.

RANUNCULUS (ra-*nun*-ku-lus)

The varied species in this genus range from weeds and wildflowers to glamorous half-hardy showpieces, but all of them, being buttercups, enjoy moist, fertile soil.

***R. aconitifolius* 'Flore Pleno', Fair Maids of France**: this very old cottage favourite grows to 60cm/2ft in sun or partial shade provided it is not allowed to dry out. The tiny, very double, pure white flowers crown stiff, widely branching stems above dark green, dissected leaves. It becomes dormant after flowering in May/June, but you should keep its position well weeded and mulched in readiness for next year.

***R. acris* 'Flore Pleno', Bachelor's Buttons**: this is the double yellow buttercup with a green centre. It flowers on almost leafless, branching stems, 75cm/2½ft tall in June. An easy plant, not invasive, but its basal clump of buttercup leaves can be divided from time to time.

R. asiaticus: this brilliant May–June flowering ranunculus looks like a small peony (10–12cm/4–5in across) and picks beautifully. It grows from a small brown tuber, and bulb merchants import different named varieties of F_1 hybrids. But none is reliably hardy, so after it has faded, in July, the tubers should be lifted, dried in the sun, and over-wintered in a frost-free place, to be planted out again, 5cm/2in deep, claws facing downwards, early in the next spring. The easy way, of course, is to buy young plants from a professional grower in April, keep them well watered, and watch them grow to 25–30cm/10–12in and flower prodigiously for about six weeks in a sunny corner of the garden, or even on a cool window-sill.

***R. ficaria*, the lesser celandine**: this has small, rounded leaves and yellow flowers, blooming March–May in well-drained, moisture-retentive soil.

R.f. 'Brazen Hussy': dark brown shining foliage offsets the yellow celandine flowers, 15cm/6in tall.
R.f. 'Salmon's White': mottled, heart-shaped leaves, pure white flowers and pale yellow buds.

R. gramineus: the leaves are not buttercup leaves, but grassy and bluish-green, in a neat clump below shining lemon-yellow buttercup flowers on wiry, branching stems, 30cm/12in tall, in May/June.

THALICTRUM (thal-*ik*-trum)

Meadow Rue

True to their family, all thalictrums like moisture. Given this, most of them are easy to grow, and you can increase your stock either by division in spring or by raising them from seed. Their leaves are as perfectly shaped and divided as a maidenhair fern's, and they fit well into narrow spaces, growing upwards, not outwards.

T. aquilegiifolium: its leaves are so like an aquilegia's that it can be mistaken for one in its small pot in a garden centre. But it grows taller (90cm/3ft), needs no staking or dead-heading, and produces tufted panicles of fluffy purple flowers in May–July.

T.a. album: white flowers.
T.a. 'Thundercloud': a fierce rosy mauve.

***T. delavayi* (syn. *T. dipterocarpum*)**: this is the desirable and difficult one, with wide-spreading panicles of tiny globe-shaped buds held high in the air on 1.5m/5ft stems in July–August. Lilac sepals fall to reveal bunches of greenish-white stamens. Each individual flowering stem should be tied to a short stake when it starts to grow and the plant should be given a sheltered place. Plant it 23cm/9in deep in good, deep, moisture-retentive soil. Even so, it may not be long-lived.

T.d. 'Album': a white form.
T.d. 'Hewitt's Double': differs from the type in having deep lilac stamens to match its sepals.

***T. flavum*, the native English meadow rue**: this is a towering (1.5–2m/5–6ft) wildflower with green leaves and fluffy yellow flowers.

T.f. glaucum (syn. *T. speciosissimum*): a Spanish variation on the theme, whose elegant leaves are a glaucous blue branching from tall, strong stems topped, in July, by fluffy sprays, 20cm/8in long, of acid-green and yellow flowers. It will look good anywhere, in any ordinary garden soil – at the back of a border or among wild-looking rushes by a pond. Tall (nearly 2m/6ft) but gentle and ready to complement its neighbours.

TROLLIUS (*trol*-le-us)

Globe Flower

Keep all trollius moist; make sure there is compost in the planting hole, mulch and water in dry weather, and it will be no trouble, flowering in sun or shade. It is excellent for picking; it does not need staking, though some varieties grow almost 1m/3ft tall, it does not spread unduly, and it is hardy. Dead-head it after its June flowering and it may reward you with an autumn repeat. It makes a neat clump of fresh green, sharply cut, pedate leaves and it is very easy to divide – either after flowering or in the spring.

T. asiaticus: this is 30cm/12in tall with dark leaves and many-petalled, open flowers of deep yellow with orange stamens.

T. europaeus: the simple lemon-yellow globe flower, native to Britain, is as lovely as any of the recent cultivars.

T.e. 'Superbus' is a specially good form, 60cm/2ft tall, with sulphur-yellow flowers in May and June.

T. × cultorum: hybrids between *T. europaeus* and *T. asiaticus* and *T. chinensis*, covering many named varieties which are all about 60–75cm/2–2½ft tall and flower in May–June; some are yellow and some are orange.

ORANGES

'Fire Globe': very deep orange, strong growing.

'Orange Princess': brilliant orange, tolerates some dryness.

YELLOWS

'Alabaster': ivory-primrose, not a strong grower, but an irresistible colour.

'Canary Bird': bright lemon yellow.

'Goldquelle': large flowers, pure yellow.

'Lemon Queen': extra free-flowering.

Trollius europaeus has up to ten beautifully incurved sepals cupped to contain its thin, pointed petals and numerous stamens.

List of Suppliers

U.K. PLANT SUPPLIERS

Herbaceous Plants

Blooms of Bressingham
Diss, Norfolk IP22 2AB

Four Seasons
Forncett St Mary, Norwich, Norfolk NR16 1JT (Mail order only)

Goldbrook Plants
Hoxne, Eye, Suffolk IP21 5AN

The Hannays of Bath
Sydney Wharf Nursery, Bathwick, Bath, Avon BA2 4ES (No mail order)

Hopleys Plants Ltd
High Street, Much Hadham, Herts SG10 6BU

Langthorns Plantery
High Cross Lane West, Little Canfield, Dunmow, Essex CM6 ITD (No mail order)

Paradise Centre
Twinstead Road, Lamarsh, Bures, Suffolk CO8 5EX

Unusual Plants
Beth Chatto Gardens, Elmstead Market, Colchester, Essex CO7 7DB

Anemones

Rupert Bowlby
Gatton, Reigate, Surrey RH2 0TA

Broadleigh Gardens
Bishops Hull, Taunton, Somerset TA4 1AE

Potterton & Martin
The Cottage Nursery, Moortown Road, Nettleton, Caistor, Lincs LN7 6HX

Clematis

Great Dixter Nurseries
Northiam, Rye, E. Sussex TN31 6PH

M. Oviatt-Ham
Ely House, 15 Green Street, Willingham, Cambridge CB4 5JA

Thorncroft Clematis Nursery
Thorncroft, The Lings, Reymerston, Norwich NR9 4QG (No mail order)

Treasures of Tenbury Ltd
Burford, Tenbury Wells, Worcs WR15 8HQ

The Valley Clematis Nursery
Willingham Road, Hainton, Lincoln LN3 6LN

Delphiniums

Blackmore & Langdon Ltd
Pensford, Bristol BS18 4JL

Rougham Hall Nurseries
(RHN Ltd), Ipswich Road, Rougham, Bury St Edmunds, Suffolk IP30 9LZ

Hellebores

Blackthorn Nursery
Kilmeston, Alresford, Hants SO24 0NL (No mail order)

Helen Ballard
Old Country, Mathon, Malvern, Worcs WR13 5PS (Mail order only)

Washfield Nursery
Horns Road, Hawkhurst, Kent TN18 4QU

Court Farm Nurseries
Honeybourne Road, Pebworth, near Stratford-upon-Avon, Warks CV37 8XT

U.S. PLANT SUPPLIERS

W. Atlee Burpee & Co.
300 Park Ave., Warminster, PA 18974
(800) 888-1477

Carroll Gardens
P.O. Box 310
Westminster, MD 21157
(301) 848-5422

Dutch Garden
P.O. Box 200, Adelphia, NJ 07710
(908) 780-2713

Henry Field's Seed & Nursery
415 North Burnett St.
Shenandoah, IA 51602
(605) 665-4491

Forestfarm
990 Tetherow Road
Williams, OR 97544
(503) 846-6963

Greer Gardens
1280 Good Pasture Island Road
Eugene, OR 97401
(503) 686-8266

Holbrook Farm
P.O. Box 368, Fletcher, NC 28732
(704) 891-7790

J.W. Jung Seed Co.
Randolph, WI 53957
(414) 326-3123

Lamb Nurseries
101 E. Sharp Ave.
Spokane, WA 99202
(509) 328-7956

Maroushek Gardens
120 E. 11th St., Hastings, MN 55033
(612) 437-9754

Mellinger's Inc.
2310 W. South Range Road
North Lima, OH 44452
(216) 549-9861

Midwest Groundcovers
P.O. Box 748, St. Charles, IL 60174
(708) 742-1790

Milaeger's Gardens
4838 Douglas Ave., Racine, WI 53402
(414) 639-2371

J.E. Miller Nurseries, Inc.
5060 West Lake Road
Canandaigua, NY 14424
(800) 836-9630

Rice Creek Gardens, Inc.
11506 Highway 65, Blaine, MN 55434
(612) 754-8090

Arthur Steffan, Inc.
P.O. Box 184, Fairport, NY 14450
(716) 377-1665

Wayside Gardens
1 Garden Lane, Hodges, SC 29695
(800) 845-1124

Weston Nurseries
P.O. Box 186, Hopkinton, MA 01748
(508) 435-3414

White Flower Farm
Route 63, Litchfield, CT 06759
(203) 567-0801

Woodlanders
1128 Colleton Ave., Aiken, SC 29801
(803) 648-7522

General information available from:
American Horticultural Society
7931 E. Boulevard Drive
Alexandria, VA 22308
(703) 768-5700

FURTHER READING

BOTANY

Flora Europeana, Vol. I Cambridge University Press
Gerard, John *Herbal* various editions
Hickey, Michael and King, Clive *100 Families of Flowering Plants* Cambridge University Press 1981
Hutchinson, J. *The Families of Flowering Plants* Macmillan 1926 (Koeltz Science Books, U.S.)
Jeffrey, Charles *An Introduction to Plant Taxonomy* 2nd edition, Cambridge University Press 1983
Kirk, J.W.C. *A British Garden Flora* Edward Arnold 1927
Loudon, Mrs Jane *Botany for Ladies* John Murray 1842
Parkinson, John *Paradisi in Sole* reprinted Methuen and Co. 1904 (Walter J. Johnson, U.S.)
Raven, John *A Botanist's Garden* Collins 1971
Rousseau, J-J. *Botany. A study of Pure Curiosity* translated Kate Otteranger, Michael Joseph 1979

GENERAL

Bowles, E.A. *My Garden in Spring* David and Charles reprint 1972 (Theophrastus, U.S.)
Chatto, Beth *The Damp Garden* J.M. Dent and Sons 1982
Fish, Margery *Cottage Garden Flowers* Faber and Faber reprint 1985
Jekyll, Gertrude *Wood and Garden* Longmans, Green and Co. 1899 (Ayer, U.S.)
Johnson, A.T. and Smith, H.A. *Plant Names Simplified* Collingridge 1931
Lacey, Stephen *The Startling Jungle* Viking 1986 (Godine, U.S.)
Lloyd, Christopher *Hardy Perennials* Studio Vista 1966
Mathew, Brian *The Smaller Bulbs* Batsford 1987 (Trafalgar Square, U.S.)
Meikle, R.D. *Garden Flowers* Eyre and Spotteswoode 1963
Philip, Chris *The Plant Finder* edited by Tony Lord, Headmain, annually
R.H.S. Dictionary of Gardening 2nd edition edited by Patrick M. Synge, Clarendon Press
Sackville-West, V. *Even More for your Garden* Michael Joseph 1958
Stuart, David and Sutherland, James *Plants from the Past* Viking 1987
Thomas, Graham Stuart *Perennial Garden Plants or The Modern Florilegium* 3rd edition, J.M. Dent and Sons 1990 (Timber Press, U.S.)
Watson, Forbes *Flowers and Gardens* Strachan and Co. 1872

PARTICULAR GENERA

Anemones

Genders, Roy *Anemones* Faber and Faber 1956

Clematis

Fisk, Jim *The Queen of Climbers* Cassell 1989 (Sterling, U.S.)
Success with Clematis The Garden Book Club 1962
Lloyd, Christopher *Clematis* revised edition, Viking/Penguin 1989 (Capability's, U.S.)

Delphiniums

Bassett, David *Delphiniums* R.H.S. Handbook 1990
Bishop, Frank *Delphiniums* Collins 1949
Phillips, George A. *Delphiniums* revised edition, Eyre and Spottiswoode 1949

Hellebores

Mathew, Brian *Hellebores* Alpine Plant Society 1989
Rice, Graham and Strangman, Elizabeth *The Gardener's Guide to Growing Hellebores* David and Charles 1993 (Timber Press, U.S.)

INDEX

Page references in *italics* are to illustration captions

PICTURE CREDITS

The author and publishers are grateful to the following for permission to reproduce illustrations:

Eric Crichton: pp. 19, 22, 46, 63, 79, 82–3, 111
Jerry Harpur: pp. 7, 11
Jacqui Hurst: p. 71 (by kind permission of Kyle Cathie)
Andrew Lawson: pp. 2, 14–15, 23, 26–7, 27, 30, 31, 34–5, 39, 43, 47, 50–1, 54, 55, 59, 62, 66–7, 70, 74, 75, 87, 90, 91, 94–5, 98, 99, 102, 103, 107, 114, 123

The endpapers and line drawings on pp. 8, 9, 10, 12, 13, 33 and 126 are reproduced by kind permission of the Royal Horticultural Society Lindley Library from Fiori's *Iconographia florae Italicae*.

Helleborus viridis

Helleborus niger

Helleborus fœtidus

Helleborus lividus

Isopyrum thalictroides

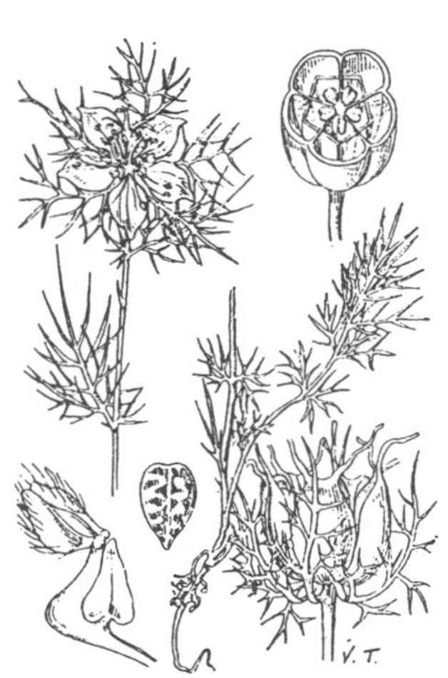
Nigella damascena

Nigella sativa

Nigella arvensis

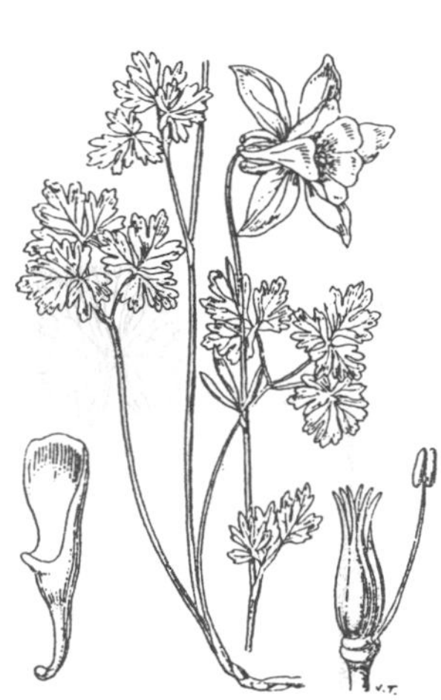
Aquilegia alpina